W9-AQO-557

Space Exploration

OPPOSING VIEWPOINTS®

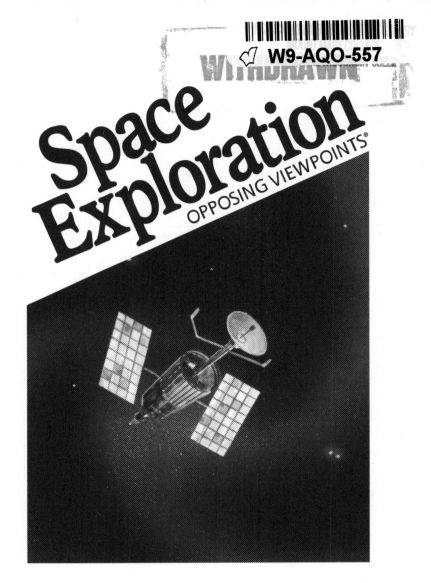

David L. Bender & Bruno Leone, *Series Editors*

Charles P. Cozic, *Book Editor*

OPPOSING VIEWPOINTS SERIES ®

Greenhaven Press, Inc.   PO Box 289009   San Diego, CA 92198-9009

Library of Congress Cataloging-in-Publication Data

Space exploration : opposing viewpoints / Charles P. Cozic, book editor.
    p.   cm. — (Opposing viewpoints series)
    Includes bibliographical references and index.
    Summary: Debates the space exploration issues of proper goals of space exploration, which programs should be pursued, the elimination of NASA, and the appropriateness of using space for warfare.
    ISBN 0-89908-197-5 (Lib. bdg. : alk. paper). — ISBN 0-89908-172-X (pbk. : alk. paper) :
    1. Astronautics—United States—Juvenile literature.   2. Outer space—Exploration—United States—Juvenile literature.   3. United States. National Aeronautics and Space Administration—Juvenile literature.   4. Astronautics, Military—United States—Juvenile literature.   [1. Outer space—Exploration.   2. Astronautics.] I. Cozic, Charles P., 1957-   . II. Series: Opposing viewpoints series (Unnumbered)
TL798.8.U5S58   1992
333.9'4—dc20   92-8149
                                                                              CIP
                                                                              AC

"Congress shall make no law . . . abridging the freedom of speech, or of the press."

First Amendment to the U.S. Constitution

The basic foundation of our democracy is the first amendment guarantee of freedom of expression. The Opposing Viewpoints Series is dedicated to the concept of this basic freedom and the idea that it is more important to practice it than to enshrine it.

Contents

Chapter 3: Should NASA Be Eliminated?

Chapter 4: Should Space Be Used for Warfare?

Why Consider Opposing Viewpoints?

"It is better to debate a question without settling it than to settle a question without debating it."

Joseph Joubert (1754-1824)

The Importance of Examining Opposing Viewpoints

The purpose of the Opposing Viewpoints Series, and this book in particular, is to present balanced, and often difficult to find, opposing points of view on complex and sensitive issues.

Probably the best way to become informed is to analyze the positions of those who are regarded as experts and well studied on issues. It is important to consider every variety of opinion in an attempt to determine the truth. Opinions from the mainstream of society should be examined. But also important are opinions that are considered radical, reactionary, or minority as well as those stigmatized by some other uncomplimentary label. An important lesson of history is the eventual acceptance of many unpopular and even despised opinions. The ideas of Socrates, Jesus, and Galileo are good examples of this.

Readers will approach this book with their own opinions on the issues debated within it. However, to have a good grasp of one's own viewpoint, it is necessary to understand the arguments of those with whom one disagrees. It can be said that those who do not completely understand their adversary's point of view do not fully understand their own.

A persuasive case for considering opposing viewpoints has been presented by John Stuart Mill in his work *On Liberty*. When examining controversial issues it may be helpful to reflect on this suggestion:

9

The only way in which a human being can make some approach to knowing the whole of a subject, is by hearing what can be said about it by persons of every variety of opinion, and studying all modes in which it can be looked at by every character of mind. No wise man ever acquired his wisdom in any mode but this.

Analyzing Sources of Information

The Opposing Viewpoints Series includes diverse materials taken from magazines, journals, books, and newspapers, as well as statements and position papers from a wide range of individuals, organizations, and governments. This broad spectrum of sources helps to develop patterns of thinking which are open to the consideration of a variety of opinions.

Pitfalls to Avoid

A pitfall to avoid in considering opposing points of view is that of regarding one's own opinion as being common sense and the most rational stance, and the point of view of others as being only opinion and naturally wrong. It may be that another's opinion is correct and one's own is in error.

Another pitfall to avoid is that of closing one's mind to the opinions of those with whom one disagrees. The best way to approach a dialogue is to make one's primary purpose that of understanding the mind and arguments of the other person and not that of enlightening him or her with one's own solutions. More can be learned by listening than speaking.

It is my hope that after reading this book the reader will have a deeper understanding of the issues debated and will appreciate the complexity of even seemingly simple issues on which good and honest people disagree. This awareness is particularly important in a democratic society such as ours where people enter into public debate to determine the common good. Those with whom one disagrees should not necessarily be regarded as enemies, but perhaps simply as people who suggest different paths to a common goal.

Developing Basic Reading and Thinking Skills

In this book, carefully edited opposing viewpoints are purposely placed back to back to create a running debate; each viewpoint is preceded by a short quotation that best expresses the author's main argument. This format instantly plunges the reader into the midst of a controversial issue and greatly aids that reader in mastering the basic skill of recognizing an author's point of view.

A number of basic skills for critical thinking are practiced in the activities that appear throughout the books in the series. Some of the skills are:

Evaluating Sources of Information. The ability to choose from among alternative sources the most reliable and accurate source in relation to a given subject.

Separating Fact from Opinion. The ability to make the basic distinction between factual statements (those that can be demonstrated or verified empirically) and statements of opinion (those that are beliefs or attitudes that cannot be proved).

Identifying Stereotypes. The ability to identify oversimplified, exaggerated descriptions (favorable or unfavorable) about people and insulting statements about racial, religious, or national groups, based upon misinformation or lack of information.

Recognizing Ethnocentrism. The ability to recognize attitudes or opinions that express the view that one's own race, culture, or group is inherently superior, or those attitudes that judge another culture or group in terms of one's own.

It is important to consider opposing viewpoints and equally important to be able to critically analyze those viewpoints. The activities in this book are designed to help the reader master these thinking skills. Statements are taken from the book's viewpoints and the reader is asked to analyze them. This technique aids the reader in developing skills that not only can be applied to the viewpoints in this book, but also to situations where opinionated spokespersons comment on controversial issues. Although the activities are helpful to the solitary reader, they are most useful when the reader can benefit from the interaction of group discussion.

Using this book and others in the series should help readers develop basic reading and thinking skills. These skills should improve the reader's ability to understand what is read. Readers should be better able to separate fact from opinion, substance from rhetoric, and become better consumers of information in our media-centered culture.

This volume of the Opposing Viewpoints Series does not advocate a particular point of view. Quite the contrary! The very nature of the book leaves it to the reader to formulate the opinions he or she finds most suitable. My purpose as publisher is to see that this is made possible by offering a wide range of viewpoints that are fairly presented.

David L. Bender
Publisher

Introduction

"Exploration of space is the challenge of our day. If we continue to put our faith in it and pursue it, it will reward us handsomely."

Wernher von Braun, rocket scientist

"I can't see voting monies to find out whether or not there is some microbe on Mars, when in fact I know there are rats in the Harlem apartments."

Ed Koch, former mayor of New York City

Twentieth-century space exploration has helped humans learn more about the universe than ever before. For example, American unmanned spacecraft, *Voyager 1* and *2*, discovered new moons surrounding Neptune and helped scientists determine the composition of Saturn's rings. Also, radar images of Venus from the *Magellan* probe in 1991 revealed evidence of earthquakes and a 4,200-mile dry riverbed, the longest known to exist in the solar system. These discoveries can help scientists better understand the nature of the galaxy and the origins of the universe.

Proponents argue that these dramatic discoveries are proof that space exploration is a profitable and worthy endeavor that benefits everyone. In addition, they cite numerous technological spinoffs from space-related research that have become everyday products and services. For example, Apollo mission technology spawned laser surgery and a more effective heart pacemaker, innovations which have improved the health of thousands of Americans. Satellite photograph technology is used in medicine to diagnose cancer and other diseases. These and other accomplishments cause many people to praise the space program, including Utah Senator Jake Garn, who once flew aboard the space shuttle *Discovery*: "For every dollar spent on space research, the private sector receives eight or nine dollars in return. What other federal program can promise that kind of return on your investment?" he asks. Garn and others believe that the rewards of space exploration far exceed its costs.

Others, however, disagree. They believe that the more than

$300 billion the United States has spent on space exploration since 1958 is extravagant and largely a waste of taxpayers' money. They point out that with today's government budget constraints, expensive space projects are an extravagance the U.S. can ill afford. "The government is no longer an automatic teller machine [for NASA]," said Howard Wolpe, chairman of a House of Representatives subcommittee that investigated NASA in 1991. These people argue that the federal government should fund programs to solve nationwide crises such as the AIDS epidemic, homelessness, poverty, and the decay of cities. Says Gar Smith, editor of the *Earth Island Journal* environmental magazine, "We need people programs rather than a space program."

Does the space program benefit Americans or waste taxpayers' money? The viewpoints in *Space Exploration: Opposing Viewpoints* address this question and others in the following chapters: What Should Be the Goal of Space Exploration? Which Space Programs Should the U.S. Pursue? Should NASA Be Eliminated? Should Space Be Used for Warfare? These issues will be of growing interest to Americans as the United States aims to expand its exploration of space in the coming decades.

What Should Be the Goal of Space Exploration?

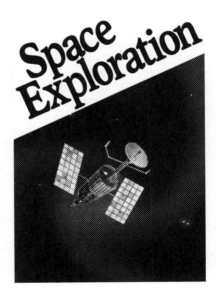

Chapter Preface

Space exploration encompasses a variety of scientific and commercial goals with many potential benefits. In the words of U.S. vice president Dan Quayle, head of the National Space Council, "Scientific ventures in space not only expand our knowledge and understanding of the universe, they lead also to development of technologies and processes that impact directly on commercial progress."

Such ventures appeal to many scientists, business leaders, and others in both government and the private sector who foresee scientific and commercial benefits from space exploration. For example, government scientists at NASA and the Department of Energy envision launching satellites that would collect a continuous supply of sunlight and transmit solar energy to earth via microwaves. Private corporations support solar satellite research because of its potential for profit. According to the Space Studies Institute's Gerard K. O'Neill, a leading researcher of solar satellite technology, "A U.S. industry in space to build power satellites for all nations would tap a world market which could exceed $250 billion a year."

However, individuals and groups striving to achieve this and other goals in space face some obstacles. Primarily through NASA, the government controls all space launches in the United States and decides which programs will receive the highest priority. Thus far, private companies have no launch facilities of their own and depend completely on the federal government to meet their launch needs. Because only a few space shuttle and other launches are made each year, companies must compete with each other to achieve their space goals. Those who favor mining the moon, for example, compete directly with those who support satellite studies of global warming.

These vying interest groups work to lobby Congress and influence space program policies in favor of their exploration goals. Which of these goals deserves the highest priority is argued by the authors in the following chapter.

> *"If we don't search [for extraterrestrial intelligence], we'll never find anything."*

The Search for Life on Other Planets Should Be a Primary Goal

Thomas R. McDonough, interviewed by Barry Karr

Both NASA and private organizations now participate in SETI, the search for extraterrestrial intelligence. In the following viewpoint, Thomas R. McDonough argues that this search is the most important goal for space exploration. He believes that extraterrestrial beings could provide earth with knowledge that would help eliminate some of humanity's problems. McDonough is an astrophysicist and the SETI coordinator for the Planetary Society, a Pasadena, California, organization that promotes space exploration. Barry Karr is the executive director of the Committee for the Scientific Investigation of Claims of the Paranormal, located in Buffalo, New York.

As you read, consider the following questions:

1. Why does McDonough believe that extraterrestrial life exists?
2. In the author's opinion, why will it be difficult to communicate with intelligent alien life?
3. How can the discovery of an extraterrestrial civilization unite nations on earth, according to McDonough?

"Searching for Extraterrestrial Intelligence," Thomas R. McDonough, interviewed by Barry Karr, *Skeptical Inquirer*, Spring 1991. Reprinted with permission.

Karr: What exactly is SETI?

McDonough: SETI stands for the "Search for Extraterrestrial Intelligence" and is the name used throughout the world for projects in which scientists are looking for objective, repeatable proof of the existence of civilizations elsewhere in the universe. Usually SETI is understood as the search for radio signals from other civilizations. . . .

Karr: Why do you think that there may be someone, or something, trying to communicate with us?

McDonough: The universe is so big that it seems very unlikely that we could be alone. The mere fact that we have several hundred billion stars in our own Milky Way galaxy, and the fact that this galaxy is just one of perhaps a hundred billion other galaxies, shows the overwhelming number of places in the universe where life could exist. And the idea that we could be alone in the universe seems unlikely just on those grounds alone. But more than that, when you look at the history of life on earth, and at what we're made of, there is nothing that seems terribly unusual when you compare the conditions on the earth with the conditions elsewhere in the universe. For example, the sun is not a weird star, it's not a rare star, it's one of the more common types of stars in the universe. Or if you look at what we're made of—carbon, hydrogen, oxygen, nitrogen—these atoms are not rare. If we were made out of platinum, for example, I might be worried about the basic ingredients for life not being widespread, but we are made out of some of the most common atoms in the universe, ones that we have detected around distant stars and in other galaxies. . . .

Extraterrestrial Signals

Karr: We actually couldn't converse with extraterrestrials if we did pick up a signal, could we?

McDonough: No, not in the sense of our saying, "Hi, how are you," and their responding, "Well, we're doing pretty good. How are you?" The problem is the enormous distances that exist in the universe. It would mean that, for example, even if they're a light-century away, a hundred light-years away, it would take a hundred years for our signal to get from here to there. And then if they replied, it would take another hundred years for that signal to get back. So it's not the kind of spirited conversation that you would like to have.

Karr: Then what would be the benefits of receiving such a signal, of hearing from another civilization?

McDonough: Well, the greatest benefit would just come from knowing that there is another civilization out there. Right now, without that knowledge, we tend to think of ourselves as being unique in the universe, and for people in one culture to think

18

Reprinted by permission of *Skeptical Inquirer.*

that they are special and very different from people in another culture. So the Russians, the Chinese, and the Americans all think of themselves as being very different from one another. But if suddenly we had positive proof that there were some two-headed little green men in another world with a more advanced civilization, we would begin to see ourselves as a single species in which our differences were trivial compared with the differences between us and them.

Karr: Why do you think that if we picked up a signal it would have to be from a more advanced civilization than our own?

McDonough: That's because the civilizations that are not as advanced as we are probably don't have technology we could detect yet. We've been on this planet for millions of years as human beings, but we've only had technology capable of spanning the distances between the stars for the past century. That's when radio was invented. That means that we're not going to be able to detect a civilization at the level of ours at Isaac Newton's time because they didn't have radio. There could be civilizations out there at that level, but we will never detect them with our present technology.

Karr: What do you think the message would be? What could they teach us?

McDonough: Oh, I think they could teach us almost everything, because a civilization that is more advanced than we are probably would have solved many of the problems we are now facing: problems of pollution, overpopulation, disease. They may have an encyclopedia that they are just broadcasting at us, which has the answers to all of our questions: how to prevent cancer, how to cure AIDS, and so forth. If we just pick up that information and can decipher it, then we may have a way of jumping ahead thousands of years in our scientific understanding and solving all of our problems.

Unsuccessful Searches

Karr: . . . What happens if in ten years we don't pick up a signal?

McDonough: I'll be very depressed. What it would mean really is that there won't be any radio signal of the kind we're looking for, but that there could be other signals out there using a technology that we haven't tried yet, or haven't discovered yet. Or it may mean that the signals are not beamed at us, or are not at the frequencies we've used. Most of the thinking so far has gone into detecting signals that are beamed like a lighthouse beacon toward us, because those kind are much easier to detect, even though they are still hard. If they're not beamed at us, if they are just alien "I Love Lucy" shows leaking off into space, then we are going to have to use a much more powerful technology that we don't yet have, and that we can't afford to develop right now. For example, we'd probably have to set up giant radiotelescopes in orbit around the earth, or better yet, on the far side of the moon.

Karr: What are the benefits of SETI if we don't ever hear from someone else?

McDonough: Well, if we don't, then the first benefit is that we become convinced that we are alone in the universe, in which case we have to be very careful of our own life. It will make us realize the unique value of our own planet in the universe. And second, it will mean that we will have searched the sky very carefully and probably will have made astronomical discoveries. One of the nice things about SETI is that as you search the sky looking for signals from other civilizations, you automatically discover other objects in the universe like pulsars, for example. . . .

The Odds of Intelligent Life

Karr: Some biologists claim that the probability of intelligence evolving is very small, since out of the billion species that have arisen here only one has achieved our level of intelligence.

McDonough: Multicellular life has only existed for half a billion years, but our sun will continue to shine without much change for another five billion years. This means that, if we hadn't evolved, billions more species would have arisen.

All it takes is for one species to succeed. Already there are many species on our planet with brains similar in size to ours: dolphins, whales, apes, elephants, and so on. If we hadn't come along, chimpanzees or gorillas might have continued to evolve larger brains. Or maybe bears or some other very different creatures would have succeeded.

As soon as a rudimentary intelligence arises—something comparable to the first tool-using primates—then the smarter ones start outwitting their less-brainy competitors. This means they can have more offspring than the dumber ones. Intelligence gives a creature the ability to defeat its competitors and to survive climate changes.

The game of life is a lottery in which you get billions of chances to play. My suspicion is that, once life evolves on a planet, intelligence has many opportunities to arise during the billions of years the average star shines. Just one success, and they own the place. If I'm right, SETI will probably succeed. If I'm wrong, it will probably fail. That's good science: a testable prediction.

Karr: How would you summarize the arguments in favor of SETI?

McDonough: Nobody on earth truly knows whether we're alone or if the universe is buzzing with life. Sticking our heads in the ground never got us anywhere in science. If we don't search, we'll never find anything. If we do search, the least we'll do is explore the universe. Let's do it!

"In undertaking space colonization, we can provide genuine hope for both our people and our planet."

The Colonization of Space Should Be a Primary Goal

Gregg Maryniak

Many scientists believe that humans will colonize the moon and nearby planets in the next century. In the following viewpoint, Gregg Maryniak argues that colonies could be constructed in space itself. Maryniak believes that these colonies would benefit the earth and its people by offering new lands for settlement. Maryniak is the executive vice president of the Space Studies Institute, a space research organization in Princeton, New Jersey.

As you read, consider the following questions:

1. In Maryniak's opinion, why will the moon be an important space colony?
2. How would space colonies sustain human life, according to Maryniak?
3. According to the author, how can space be used to provide energy to earth?

Gregg Maryniak, "How Space Colonies Could Benefit Earth," *The Christian Science Monitor*, January 2, 1992. Reprinted by permission from The Christian Science Monitor and the World Media Project.

Space colonization means much more than Antarctic-style research habitats on the moon or other planets for an elite group of astronauts. Space can be colonized and provide Earth with the equivalent of the New World that Columbus "discovered" in the 15th century.

Space colonies can supply clean energy necessary for human survival in the 21st century. In addition, they can provide new homelands and an expanded ecological niche for our species.

For many people, the term "space colony" brings to mind visions of domed cities on the moon or the surface of a hostile planet. Since September 1974, however, the words have had a very different meaning. That month's issue of *Physics Today* contained an article by Princeton University professor and nuclear physicist Gerard K. O'Neill entitled, "The Colonization of Space." Dr. O'Neill proposed construction of large-scale habitats built in free space rather than on the surface of planets.

Orbiting Colonies

Building the structures in space would allow the inhabitants to select whatever gravity level they desired by controlling the rate of rotation of the habitat. O'Neill showed that even if relatively simple materials such as steel cables were used in colony construction, habitat cylinders of up to 20 miles (32 kilometers) in length and 4 miles (6.4 kilometers) in diameter could be built to house up to 1 million people under comfortable conditions.

Early habitats would be much smaller, with populations of hundreds or thousands. Each habitat would have provisions for agriculture and closed-cycle life support so that once a colony is established, very little outside material would be required to sustain it.

To obtain construction materials for these settlements from Earth would obviously not be economical. Instead, O'Neill proposed using materials already in space.

The first source of raw materials would be the surface of the moon. Thanks to the Apollo missions and the Soviet sample return probes, we know that the required elements are present in abundance.

Because the moon has no atmosphere and only one-sixth Earth's gravity, it is possible to use an electromagnetic catapult (called a mass driver) to launch raw materials to a point in space without incurring the costs of chemical-rocket transport.

Space colonies were the subject of three studies the National Aeronautics and Space Administration conducted in 1975, 1976, and 1977. These projects examined the issues of closed-cycle life support, radiation shielding, habitat design and construction, economics and logistics, and lunar and asteroid mining.

The initial economic reason for the colonization of space

would be to use the resources of space to provide for the needs of our home planet. O'Neill proposed that the space colonists use low-cost space resources to construct large solar platforms to collect the sun's energy and convert it into electricity. This electrical power would be transmitted to Earth's surface in the form of a high-frequency radio beam. The beam would be received by a special antenna and rectifier array, which would convert it back into electricity with an efficiency of about 90 percent.

Large-Scale Colonies

Space has strong advantages over the surface of moons or of other planets as the site for location of human habitats (colonies.) In space we can build habitats that are large in scale—far larger than could be built on a planet, because those structures will not have to withstand gravity. But they can rotate, to provide for their inhabitants internal gravity equal to Earth's norm, for which we have evolved over millennia. A space habitat will be a spherical shell of metal and glass, enclosing a normal, breathable atmosphere—something that does not exist on any planet other than Earth.

Gerard K. O'Neill, *The World & I*, January 1992.

Solar-power satellites would provide electricity without the air pollution and atmospheric heating caused by burning fossil fuels. The use of space resources would enable these satellites to be built at less than 10 percent of the cost of launching construction materials from Earth.

In addition to the workers and their families, space colonies would contain many of the professions found in any small terrestrial town. Space settlements would also address human needs beyond the physical and economic. By its very nature, each habitat would have a high degree of self-sufficiency and independence.

Independence in space colonies need not, however, mean isolation. Communications between Earth and space colonies, or from one colony to another, would be a relatively simple matter with thousands of communities within a few light-seconds of each other. Although transport from Earth's surface to a colony or colony group is likely to remain relatively expensive, the cost of travel from one free-space habitat to another can be very small.

Ultimately, space colonies could be built anywhere in the solar system. By increasing the size of the mirrors used to direct sunlight into the living and agricultural sections, it would be possible to support habitats beyond the orbit of Pluto if we so desire. Given the known resources of the asteroids, there is sufficient material to construct habitats capable of supporting populations thousands of times larger than that of Earth. By increasing our ecological niche to include the solar system, the human species would become much less likely to be destroyed by a single natural or man-made catastrophe.

Since 1977, most of the research on space colonies and related technologies has been carried out by the nonprofit Space Studies Institute (SSI) of Princeton, N.J. SSI conducts research and serves as an information resource to industry, government, and academia.

How Space Colonies Work

A space colony has three essential characteristics. First, a colony must have a tightly closed-cycle ecological system capable of replenishing the colonists' air, water, and food with only trace elements required from outside the system. Second, it must have enough radiation-shielding so that indefinite stays are possible. Third, it must provide sufficient artificial gravity to permit the inhabitants to reside on a permanent basis without bone-calcium loss or other harmful effects of prolonged exposure to microgravity.

In addition to this description of technical characteristics, to be viable the colony would need an economic basis as would any successful town on Earth.

We know quite a lot in general about how to provide radiation-shielding for a space habitat. The secret is to have sufficient mass to stop the high-energy particles collectively known as cosmic rays, as well as radiation from the sun.

At present, space stations such as Skylab, Salyut, and Mir have flown in such low orbits that Earth's magnetic field has provided considerable protection. During the Apollo flights to the moon, it was physically impossible to provide shielding, so the astronauts' exposure to cosmic rays was limited by time. When we have the capability to process lunar soil for construction materials and propellants, we will have slag and other "waste" products that can be used as a radiation barrier. It is also theoretically possible to use superconducting magnets to provide an artificial magnetic field. However, it is unlikely that such a system would be as reliable as a simple physical shield.

Life-Support Systems

Research groups in the United States and the former Soviet Union are advancing the state of the art in Closed Environmental Life Support Systems (CELSS).

For many years, the Soviets were the undisputed leaders in CELSS research, with humans involved in systems experiments in large-scale facilities such as the Bios-3 experimental chamber in Krasnoyarsk, Russia. A privately funded US organization called Space Biospheres Ventures has surpassed the scale of the Soviet experiments by constructing a facility in the desert outside Tucson, Ariz. This facility is supporting eight people in near-complete material closure for a planned two-year period.

Space Colony Would Have Cheap Source of Energy

In our society, materials are relatively cheap and energy is relatively expensive. However, a space colony would be quite the reverse. Their energy would be very cheap and abundant—and totally reliable, coming from the sun.

Gerard K. O'Neill, *The Christian Science Monitor*, January 2, 1992.

This large facility, called Biosphere II, encloses about three acres and contains several different climate zones, including a tropical rain forest and an intensive agricultural area. These projects and work underway at the Kennedy Space Center, the Disney Epcot Center's land pavilion, and the Environmental Research Laboratory of the University of Arizona indicate that it will be feasible to sustain relatively complex closed-cycle ecological systems beyond Earth's atmosphere.

The Gravity Question

Since Isaac Newton, scientists have known how to provide "artificial gravity" by rotation. For purposes of space colony design, however, we still lack the answer to one fundamental question: How much gravity does it take to maintain normal human functions in space?

According to many scientists, there is a threshold level of gravity that will trigger normal physiological response. At present, we simply have no data to determine what level of gravity is necessary. In 1986 President Reagan's National Commission on Space recommended that the US develop a variable-gravity research facility that could provide the needed information.

During the NASA space-colony design workshops, participants assumed that Earth-normal gravity would be required in some locations. If, however, one assumes that a human needs only one-sixth of terrestrial gravity, or even less, one could construct small-scale space colonies in the very near future from such raw materials as empty space-shuttle external tanks.

Technological Progress

Much progress is being made in technologies essential for the construction of space colonies. For example, prototype mass-driver electromagnetic launchers have been developed at the

Massachusetts Institute of Technology and Princeton University. Accelerations have gone from 33 gravities to more than 1,800 gravities. This means that the length of the accelerator on the moon can be reduced from 8,900 meters (about 5 miles) to 160 meters (about 500 feet). Physical- and chemical-processing techniques to obtain propellants and construction materials from lunar soil have been demonstrated at laboratory scale.

Studies by General Dynamics and SSI have shown that most of the mass of solar-power satellites can come from lunar materials. SSI's work shows that the cost of a lunar-sourced power satellite is only 3 percent of the cost of the same satellite built from Earth-launched materials.

In 1988, NASA conducted what it called the Lunar Energy Enterprise Study in cooperation with representatives of electric-power utilities and other industries. This study recommended further examination of lunar-sourced solar-power satellites.

Benefits to Earth

In 1991, a workshop conducted by the International Astronautical Federation at a conference on solar-power satellites suggested a vigorous program of international experimentation on the use of space resources to provide energy to Earth. Although we have general information about the composition of the moon and asteroids, one of the largest opportunities in the development of space is the search for specific resource sites.

Today there is a growing realization that the world's space programs must generate real value for their constituents. Providing clean energy for Earth and new lands for settlement and exploration can take space exploration and development beyond the "flags and footprints" missions of the 1960s. In undertaking space colonization, we can provide genuine hope for both our people and our planet.

"We can process natural resources on the Moon and Mars into products we need."

Exploiting the Resources of Other Planets Should Be a Priority

Thomas A. Sullivan and David S. McKay

Resources from the moon and Mars are necessary for space colonization and exploration, argue Thomas A. Sullivan and David S. McKay in the following viewpoint. Sullivan and McKay contend that water, oxygen, carbon dioxide, and metals can be obtained in space to build outposts, propel spacecraft, and grow food, thus avoiding the costs and difficulties of transporting such supplies from earth. The authors also state that solar energy and helium collected at the lunar surface can provide earth with environmentally safe energy. Sullivan and McKay are scientists in the Solar Systems Exploration Division at the NASA Johnson Space Center in Houston, Texas.

As you read, consider the following questions:

1. How will the use of space resources result in more efficient launches, according to Sullivan and McKay?
2. Why do the authors believe it will be difficult to extract water from Mars?
3. According to the authors, how can robots assist in space resource development?

Adapted from *Using Space Resources* by Thomas A. Sullivan and David S. McKay. Washington, DC: U.S. Government Printing Office, 1991.

There is every indication that exploring the Solar System and expanding humanity's permanent presence to the Moon and Mars will provide significant benefits to America. Over the past three decades, America's space program has more than paid back its cost in a number of ways, both tangible and intangible. Even though it will continue to be an expensive undertaking, a far higher price would be borne by our nation in avoiding such a mission. However, the importance, the enormity, and the inevitability of the program require immediate focus on ways to lower its cost without reducing its scope or increasing its risk. What can we do to lower the price tag of this ambitious exploration program while still accepting its challenge?

Using Space Resources

We can process natural resources on the Moon and Mars into products we need at an outpost, thus avoiding the need to bring them from Earth. These resources can provide us with oxygen and nitrogen to breathe and water to drink. We can produce propellants for our spacecraft. Carbon dioxide (CO_2) can be extracted from the lunar soil, or regolith, to support plant growth for food. We will learn to fabricate bricks and panels from local materials and use them for constructing habitats, workshops, and storage buildings. We will extract metals from local rocks and soil to make beams, wires, and perhaps even solar power cells. In short, many of the essential materials needed for life on the new frontier can be produced from local resources.

This program is ready to move from the concept stage into research and development. Indigenous Space Materials Utilization (ISMU—formerly called In-Situ Resource Utilization) can provide a reduction in cost and can increase our capabilities significantly as we develop and expand a lunar or Mars outpost. Our goal for the ISMU program is to free these outposts from total reliance on the Earth as soon as possible, thereby rapidly shifting the nature of our space transported cargo away from bulk materials, such as propellants and building materials, to additional people and complex equipment.

Liquid Oxygen

As currently envisioned, a large part of the cost of a lunar outpost will be that of bringing supplies and propellant from Earth. The major component of this propellant is liquid oxygen (LOX). NASA's *Report on the 90-Day Study on the Human Exploration of the Moon and Mars* estimated that the amount of mass launched to low Earth orbit (LEO) could be reduced by 300 tons per year if LOX were produced on the Moon. This is equivalent to 10 Shuttle launches at a cost of several billion dollars per year. Even if we develop a new heavy lift launch vehicle (HLLV), the

29

price tag to launch this propellant will be huge. By producing the oxygen on the planet where it is actually needed, there is no shipping penalty. Thus, the amount required is much smaller.

Abundant Space Resources

Space is a unique store of resources: solar energy in unlimited amounts, materials in vast quantities from the surfaces of the Moon and Mars, gases from the Martian atmosphere, and the vacuum and zero gravity of space itself. With suitable processing, these raw resources are transformed into useful products. These products, while increasing exploration efforts, provide bulky materials at a fraction of the cost of transporting huge masses from the deep gravity well of Earth.

The Synthesis Group, *America at the Threshold*, 1991.

The use of lunar-derived LOX is thus a high-leverage item because it frees our space vehicles from the inefficient and costly exercise of shipping bulk propellants. The total cargo that must be shipped to the Moon will be reduced significantly. And for the remaining flights, instead of transporting large quantities of LOX to the Moon, more people, complex equipment, and scientific instruments can be shipped to provide additional capabilities at the lunar outpost. This effect can be increased dramatically when we also produce lunar-derived fuels, such as hydrogen or methane.

Perhaps as important as propellant production will be the use of the regolith for the manufacture of basic material. While it is true that much of the cargo arriving on the Moon will be extremely complex equipment, there is a real need for simple, basic infrastructure such as roads, rocket blast protection, and structures for habitats, storage, and equipment repair. If brought from Earth, the mass required for these uses would be enormous. For example, just for protection from solar particle radiation, the amount of mass that would have to be brought to the Moon represents several Space Shuttle launches. Based on present launch costs, the expense of transporting the hundreds of metric tons needed to protect an early habitat from this dangerous radiation would surpass a billion dollars.

Just as important as cost is the high flight rate which would be needed to support this effort. This would strain the capacity of our launch systems. When LOX is produced on the Moon, fewer flights to LEO will be necessary. On-orbit assembly and processing at Space Station Freedom (SSF) can also be reduced. Indeed, the additional facilities that would otherwise be required here on Earth and at SSF represent further cost savings made possible through an ISMU program. . . .

There may be an opportunity for lunar resources to play a role in the energy industry here on Earth. Power generation is a vast and growing market. Energy is a product that may legitimately be worth bringing back to the Earth's surface from the Moon.

Future Energy Needs

How will we do this? In 1989, a NASA report concluded that, for the energy needs of the next century, we need to consider two alternatives enabled by a lunar outpost: solar energy collected on the lunar surface and beamed back to Earth via microwaves, and the return to Earth of a light isotope of helium, He-3. Both of these options would largely avoid the biggest problems of energy generation here on Earth: pollution, acid rain, ozone generation, carbon dioxide production with its potential for global warming, and large operations with highly radioactive fuels.

Along with the other light gases mentioned above, we can extract an isotope of helium from the regolith. He-3 has the potential to be used in fusion reactors here on Earth to provide electrical power. This technology, while still in the research stage, promises to be much "cleaner" than current, *fission*-based plants which consume uranium, and even cleaner than those fusion plants currently under development which would use radioactive tritium as a fuel. Why would we go to the moon for this material?

- It is nearly absent from Earth as a natural resource.
- Millions of kilograms of it are present in lunar soil, albeit at very low concentrations.
- It may be the fuel for the next generation of electric power plants, providing nearly pollution-free power.
- It may, in the 21st century, replace our dwindling supplies of fossil fuels.
- It may be worth $2,000,000/kg.

Along with deuterium, which can be extracted from sea water, He-3 is the primary fuel of a clean nuclear fusion reactor currently being investigated by U.S., European, and Japanese fusion research scientists. Some of these scientists believe that a demonstration fusion power reactor using the He-3/deuterium reaction can be built within 10 or 15 years and a commercial power reactor within 20 years. It would generate only a very slight amount of radioactivity, equivalent in nature to that produced by hospitals in their nuclear medicine areas. When used in this plant, He-3 would have so much energy that it would require only 20 tons—less than one Shuttle load—to supply all the electricity used in the United States in a year. The current cost of fuel used to provide this electricity is tens of billions of dollars—and going up. We can estimate that the single Shuttle-load of He-3 might be worth about this same amount—or more when the environmental impact of fossil fuels is included. . . .

Mars Resources

Upon reaching Mars, we again have a world with resources that can be used to expand our capabilities. The martian atmosphere, consisting mostly of carbon dioxide, can be processed to release oxygen for life support or propellant use. Carbon monoxide, which could be a moderate performance rocket fuel, is the coproduct. By combining this oxygen with a small amount of hydrogen, water for a variety of uses could be produced for only a fraction of its mass if brought from Earth. One good aspect of atmosphere utilization is that no mining is involved. Simple gas handling equipment can be used, providing a much more reliable system.

Reducing Exploration Costs

The idea of "mining" resources on another planet to support operations there, plus the return trip to Earth, is an innovation. By using these so-called *in situ* resources, we can greatly reduce the mass that must be lifted out of Earth's gravity well. With less mass to launch, we can drastically lessen transport costs for settling the Moon and exploring Mars.

Kumar Ramohalli, *The Planetary Report*, January/February 1991.

Life support technologies routinely deal with the conversion of CO_2 to other compounds, including methane. This process was discovered nearly one hundred years ago and is still used in many chemical plants today. A direct application of this technology to the martian atmosphere would allow for the production of oxygen, methane, and water by bringing only a small amount of hydrogen. Thus, large quantities of propellant could be leveraged from minimal import mass. As described earlier, a rocket engine using methane and oxygen could be developed for use in both lunar and martian spacecraft. This could enable another large cost savings for the SEI [Space Exploration Initiative] by utilizing those materials available at the Moon or Mars.

Planetary scientists agree that water is available at the poles of Mars in the form of ice. It is likely, but not certain, that water is available elsewhere on the planet, perhaps as a permafrost layer or bound as a mineral hydrate. If the robotic missions in the early stages of the SEI provide evidence of water, there is every reason to believe that a process can be developed to make it available for human use. It is likely that one could even extract enough water to produce both hydrogen and oxygen propellant for the launch back to orbit and even the return trip to Earth, thus reducing the size of the spacecraft leaving Earth for Mars. Telerobotic mining at distances as far as Mars is not practical, however, and totally automated systems would need to be developed. And, at the more accessible latitudes near the equator, any water is likely to be found at a lower depth, compounding the problem.

The two moons of Mars, Phobos and Deimos, may also be rich in water. Processing at the extremely low gravity present on these bodies will require some innovative equipment. While early exploration scenarios suggest it would be difficult to bring this promise to fruition, future operations on or near Mars could easily make use of the potential within these bodies. Many asteroids are believed to be of similar composition and are also likely targets for utilization once we have honed our ability to operate highly complex equipment at distances so remote that teleoperation is not feasible. For the near term, however, the SEI requires the development of an ISMU program which focuses on the Moon and Mars. . . .

A Course of Action

We have in hand a good understanding of the opportunities where an ISMU program can provide vast savings to the SEI. Without such a plan, it is uncertain that America would choose to bear the expense of such a push into the solar system. The specific products outlined in this viewpoint define the areas where we must develop further technology to bring these promises to fruition.

What do we need to do next?

- We must rapidly develop and expand the technologies that would make possible the early use of available resources when we send people to the Moon and Mars.
- We must develop small, early robotic ISMU experiments to send before we send people. These systems will demonstrate the production of useful products in the actual environments of the Moon and Mars. We will learn from them, and they will influence the design of the production plants which will follow. This is a low risk approach.
- We must include strategies that aim toward maximum self-sufficiency as major goals of the SEI. This should include designing other elements of the outposts and spacecraft to maximize these benefits.
- We must implement the recommendations of the National Commission on Space and the Ride Report, which include

continuing research and technology development programs focusing on processing indigenous raw materials into propellants and construction materials.

Only by learning to live off the land will we be able to take the next "giant leap for mankind" and continue our journey to Mars and beyond.

4 VIEWPOINT

"The space science program warrants highest priority for funding."

Scientific Research Should Be a Priority

The Augustine Committee

The Augustine Committee, formally known as the Advisory Committee on the Future of the U.S. Space Program, was a temporary government panel formed in 1990 to examine the state of the U.S. space program. In the following viewpoint, the committee argues that the goal of space exploration should be science, because knowledge and discoveries in space are the most significant products of the space program. The committee maintains that U.S. space research, which has increased humanity's understanding of the universe, must remain strong to further space projects.

As you read, consider the following questions:

1. How has the United States increased its space science expertise, according to the authors?
2. Why does the committee believe that it is important to balance both large and small space research projects?
3. According to the authors, how can university research benefit NASA space projects?

Adapted from *Report of the Advisory Committee on the Future of the U.S. Space Program* by the Augustine Committee, December 1990.

The United States' civil space program was rather hurriedly formulated some three decades ago on the heels of the successful launch of the Soviet Sputnik. A dozen humans have been placed on the Moon and safely returned to Earth, seven of the other eight planets have been viewed at close range, including the soft landing of two robot spacecraft on Mars, and a variety of significant astronomical and other scientific observations have been accomplished. . . .

The material foundation of any major space project is its "technological base." It is this base that produces the key building blocks, or "enablers," that make major missions possible— new materials, electronics, engines and the like. The technology base of NASA has now been starved for well over a decade and must be rebuilt if a sound underpinning is to be regained for future space missions. . . .

It is our belief that the space science program warrants highest priority for funding. It, in our judgment, ranks above space stations, aerospace planes, manned missions to the planets, and many other major pursuits which often receive greater visibility. It is this endeavor in science that enables basic discovery and understanding, that uncovers the fundamental knowledge of our own planet to improve the quality of life for all people on Earth, and that stimulates the education of the scientists needed for the future. Science gives vision, imagination, and direction to the space program, and as such should be vigorously protected and permitted to grow, holding at or somewhat above its present fraction of NASA's budget even as the overall space budget grows. . . .

Scientific Gains

American scientists and engineers have used opportunities for access to space to advance human understanding of ourselves, our planet, our solar system, and our universe—from the discovery of the Van Allen belts to the establishment of X-ray astronomy, from the high resolution photos of the planets, their satellites, and rings to the global weather monitoring and forecasting system, from the growth in a microgravity environment of very large crystals to the age-dating of the Moon with lunar samples, from the detailed mapping of the Earth's polar ozone depletions to the precise measurement of the "Big Bang" residual radiation, from the discovery of the effects of microgravity on bone growth and healing in mammals to direct measurements of million-degree solar system plasmas, and from the discovery of the enigmatic, rare repeating gamma ray bursters to the finding of ancient and active volcanoes on other planets and satellites. These achievements and the understanding gained from them will continue to be one of the most significant products of the nation's investment in the civil space program. The cost of this effort, in recent years, has been on the order of 20 percent of NASA's budget.

© Boileau/Rothco. Reprinted with permission.

With so spectacular a set of achievements as a foundation, and with a substantial number of space projects underway, the U.S. space research enterprise should be healthy and flourishing. Yet discussions with researchers within NASA and in the university community reveal that there is significant discontent and unease about what the future may hold for U.S. space research. The reasons for these concerns have been documented in some detail in the 1986 report entitled "The Crisis in Space and Earth Science" issued by the NASA Advisory Council. They include such factors as (a) the widening of research horizons in response to past accomplishments so that there are now more opportunities than can be accommodated by the available resources; (b) the space technology required to support new advances is often more costly and sophisticated than in the past; (c) the growing

37

complexity of interactions between NASA and its larger and more diverse research community; and (d) program stretch-outs, delays and cancellations that waste creative researchers' time, squander resources, and decrease flight opportunities. We believe that many of these reasons continue to exist.

Research Is the First Priority

An underlying basis for the concern of the research community has been that the strategies, goals, objectives, and programmatic requirements of the research program have not been adequately distinguished from the parallel national objective of placing humans in space.

Mechanisms are needed which alleviate the more serious of these problems so that the talents and capabilities of America's space researchers, both inside and outside of NASA, can be focused on substantive future opportunities. We strongly affirm the central role of research in the U.S. civil space program, hence:

Recommendation 1: That the civil space science program should have first priority for NASA resources, and continue to be funded at approximately the same percentage of the NASA budget as at present (about 20 percent).

We note that this recommendation carries with it the responsibility for the research community and NASA to use these resources in a prudent manner to carry out pioneering research. To do this, the research community must understand and appreciate, as well as participate in, the planning and budgetary process. To facilitate execution of this recommendation, we propose:

Recommendation 2: That, with respect to program content, the existing strategic plan for science and applications research proposed by NASA with input from the science community be funded and executed.

Let Smaller Projects Flourish

The present strategic plan provides appropriate balance to the research program that must be maintained across the disciplines, as well as across the methodologies for carrying out the research. In particular, an appropriate mix must be achieved among small, medium, and large projects. A trend toward the development of large projects has developed in recent years, driven by several factors. These include the natural evolution in requirements of some research fields and the "new star" process employed by NASA, the Office of Management and Budget and the Congress for initiating projects to carry out research. This latter process sometimes encourages a "piling-on" of research objectives, as well as of researchers, in order to strengthen fiscal justification. An environment needs to be created that will encourage small, fast-paced projects as well as large projects and enable both to flourish.

Research support activities, such as mission operations and data analysis programs, as well as many portions of the advanced technology development program, represent the life blood of civil space research. These activities, together with sub-orbital balloon and rocket projects, are the centerpiece of university professor and student involvement with the civil space program. Such activities encourage substantial numbers of scientists and engineers, beyond those involved in hardware development for major space flight projects, to participate constructively and creatively in the space program.

"*Access to space has created . . . the ability to observe our planet holistically, learn how it works and perhaps control its future.*"

Exploration Should Be Used to Protect Earth's Environment

John A. Dutton

In a 1987 NASA study, former astronaut Sally Ride proposed the Mission to Planet Earth, a program that calls for outer space to be used to improve earth's environment. In the following viewpoint, John A. Dutton agrees and argues that space laboratories and satellites should be developed to analyze environmental conditions on earth. Dutton states that such observations from space will improve scientists' understanding of global warming and other threats to the environment. Dutton is dean of the College of Earth and Mineral Sciences at Pennsylvania State University in State College, Pennsylvania.

As you read, consider the following questions:

1. According to Dutton, how does human population affect the climate?
2. Why does the author believe that the use of computers is critical to improving the environment?
3. In addition to the environment, how can observing earth from space be beneficial, according to Dutton?

John A. Dutton, "Mission to Planet Earth," *The Planetary Report*, January/February 1990. Reprinted with permission.

Access to space has created astounding possibilities for the residents of planet Earth: the ability to survey nearby planets and perhaps someday walk upon them, as we have already on the Moon; the ability to look back through time toward the origin of the universe; and the ability to observe our planet holistically, learn how it works and perhaps control its future.

Mission to Planet Earth

The missions to explore other planets and the mysteries of the universe are among our greatest achievements, stimulating our intellectual and technological capabilities. And the many missions we have launched to observe Earth have taught us much. But we have not yet learned enough to really understand how our planet works and what its future might be. That future may be in doubt, as a result of human activities that are altering natural balances. We need to understand this planet and the changes now in progress. We need to take another giant step for humankind—a Mission to Planet Earth.

Plans for such a mission . . . have been developed by NASA in cooperation with other nations. The first mission phase will create the Earth Observing System (*Eos*), consisting initially of five polar-orbiting platforms with instruments to observe Earth's surface and atmosphere. The first platform will be launched in 1997, according to present plans. In the second phase of Mission to Planet Earth, the polar orbiters will be complemented by equator-orbiting platforms; these will be in geosynchronous orbit, matching Earth's rotation and thus providing a continuous view of an entire hemisphere.

These satellite systems will yield an unprecedented flow of information about the planet and the interactive processes that control its evolution. But obtaining these data from space is only the first step.

Earth System Science

Doubling in number every 50 years, the human species is changing the delicate physical, chemical and biological balances that shape the climate of the planet. The ozone layer, which absorbs ultraviolet radiation and thereby permits life to exist on land, is being depleted, most dramatically at the South Pole, by human use of chlorofluorocarbons in refrigeration and industrial processes. Use of fossil fuels and tropical deforestation are increasing the carbon dioxide content of the atmosphere. If this process continues, the greenhouse effect may lead to global warming, melting polar ice and rising seas; resulting changes in local climatic conditions would alter the distribution of vegetation and agricultural productivity.

First of all, we need to document the changes that are occurring and ascertain the rate of change. We'll have to learn to recognize which changes are driven by human activities and which are results of natural variations. The only way to obtain this level of understanding is to observe the entire Earth from space for several decades.

Monitoring the Environment

The space age requires a global vision to meet new challenges—from threats to the environment to the desire to explore new frontiers. . . .

We will be relying on the expertise of scientists and other scholars from space agencies around the world to help us lay the groundwork for the exploration of space and in the proposed monitoring of the global environment. However, it is my hope that the general public will also share in the excitement of Mission to Planet Earth and get involved in this opportunity to increase our understanding of our planet.

George Bush, *Final Frontier*, January/February 1992.

To understand how the whole Earth works as an integrated system, we must transcend the present specialization of disciplines to create a new science—an Earth system science—that can identify and comprehend the interactions among the atmosphere, oceans, land surface, life on land and in the sea, and the sheets of ice on land and sea. We know today that the chains of cause and effect are wondrously delicate and complex. The physical environment affects biological processes and in turn biological processes, interacting with the chemistry of the atmosphere and sea, modify the physical environment, most notably the radiation balances of the planet.

Finally, we must learn to predict reliably what conditions on planet Earth will be like decades and centuries in the future. To preserve Earth, and ourselves, we must foresee what will happen if we continue on our present course and, in contrast, what would happen as a result of specific measures in the areas of energy, pollution, reforestation and human population growth.

Such predictions, taking into account the interactive complexity of the entire Earth system, can only be made with computer models that simulate the behavior of Earth. Today computer models that combine the known physics of the atmosphere with data on current conditions can predict weather conditions a few days or a week ahead. Tomorrow we'll have models that resolve the interactions of the Earth system—including physical, chemical and biological subsystems—and these models will enable scientists to predict, at least on a statistical basis, what global and regional environments will be decades, and maybe centuries, into the future.

To develop models that will accurately simulate the workings of the Earth system, we must learn the details of how its many processes interact, from local *and* global perspectives. Mission to Planet Earth, beginning with *Eos*, is designed to produce that information—the magic mirror, if you will, to let us foresee the future of planet earth.

The Plan for *Eos*

Eos will be the largest single scientific initiative ever undertaken by NASA, or for that matter by the peoples of the world. The project will have three components: the instruments in space, an information system to distribute the results of observations and a program of interdisciplinary research focusing on application of the data to key issues in Earth system science (including the global water cycle, controls on biological productivity and atmospheric chemical cycles).

The five *Eos* polar-orbiting platforms—two each to be built by NASA and the European Space Agency and one by Japan—will carry 8 to 12 instruments per platform. The instruments, each with unique and advanced capabilities, are being designed by separate groups. Another 28 interdisciplinary research groups have already been funded to begin planning for the effective use of the data. Today, scientists and engineers from 168 institutions, universities and national laboratories in the US and 12 other countries are creating the *Eos* instruments and interdisciplinary centers. The *Eos* program will cost more than $1 billion per year over its lifetime. By way of comparison, this cost is only a little more than the Department of Defense spends each day of the year.

The NASA *Eos* polar-orbiting platforms will be large spacecraft, each approximately 12 by 4.3 meters (36 by 13 feet), weighing 12,000 kilograms (30,000 pounds). NASA will launch them from Vandenberg Air Force Base aboard *Titan IV* rockets into Sun-synchronous orbits (for continuous lighting). At an altitude of 705 kilometers (441 miles), the *Eos* satellites will obtain nearly global coverage every two days.

Earth Data and Patterns

The design of *Eos* instruments reflects an emphasis on long-term continuity of observations and on simultaneity of observations by different instruments on the same platform. Simultaneity lends depth to our "picture" with extra wavelengths and increases accuracy since the data from one instrument will often be used to correct or adjust data from another.

The *Eos* instruments in space will produce a veritable flood of

43

information about Earth. The two NASA platforms will send data at an average rate of 70 megabits per second, but at peak mode will produce data at the rate of some 300 megabits per second. To accommodate these unprecedented flows of information, NASA is designing the *Eos* Data and information System (Eos-DIS), which will make results available to scientists throughout the world. The successful design and operation of Eos-DIS is an even greater challenge than creating the instruments to fly in space. Nevertheless, scientific planners and NASA are committed to *Eos* as an information system about Earth and its processes, above and beyond the space mission component.

Observations from space reveal patterns that sweep across the face of Earth. Most familiar are the cloud patterns seen by weather satellites, but there are other intermingling patterns—in the distribution of temperature and wind at the sea surface, vegetation and vegetative stress, biological productivity in the ocean, varieties of surface minerals and the ridging of polar ice.

These tantalizing patterns, in their variety, reflect the complexity of the Earth system. They are our key to unlocking the secrets of change. We need to convert these patterns into quantitative information on processes and their rates.

For example, in cloud patterns we can recognize atmospheric processes on two scales. On the larger scale, cloud patterns reveal wind direction and speed, they indicate where convection is intense and they provide information about the distribution of water vapor. From cloud patterns we can discern and predict the regular behavior of frontal systems. On the smaller scale, cloud patterns reveal internal processes that involve vertical motion, condensation of water vapor, and absorption and emission of radiation. Thus with clouds we have some experience in relating patterns to dynamics. The same conversion from pattern to process must be effected for all the elements of the Earth system.

Soil, Water, and Air

Even what we call the *solid* Earth is dynamic. The location of mountain chains and rifts in the sea floor, along with variations in Earth's gravitational and magnetic fields, combine to reveal the effects of convection in the mantle and the resulting motion of the tectonic plates that make up the crust. From these patterns we can infer processes at work deep within Earth.

Paradoxically, we don't know quite as much about certain processes taking place at the land surface—specifically, in the interactions of soil, vegetation and atmosphere. Observing the radiation emitted or reflected from the land surface, we can learn about its "heat budget," the sum of all transfers of energy between the atmosphere and the land and vegetation. While radiative components of this budget can be measured directly from space, other components, such as the evaporation rate, will have to be inferred, perhaps from remote observations of surface temperature and soil moisture content.

Our Fragile Planet

The planet we live on is much more fragile than we thought it was. Not that it's going to break apart, but it's very sensitive to the changes that take place on its surface and even in the interior. And we can affect those changes; in fact, we're the cause of some of them. So it's important to understand the earth's ecosystem on a global scale and how it's changing. The only way we can effectively do that is from space, because that's the only place where we can get a view of the entire planet.

Sally K. Ride, *Ms.*, July/August 1987.

Much more difficult to ascertain are the processes involved in the balance and evolution of vegetative cover. Vegetation interacts with both the soil and the atmosphere, using nutrients from above and below the surface, and in the process alters the chemical composition of both land and air. However, we don't as yet know how to resolve that vast and complicated exchange of elements and energy.

It's critical that we learn to put together local records with sufficient detail *and* global records over the long term in order to get the whole picture of the chemistry occurring at Earth's surface. The vegetation on land and the flora and fauna of the oceans exchange chemical constituents with the atmosphere, soil and oceans, creating biogeochemical cycles. These biogeochemical cycles are important to us because they control the concentration of gases in the atmosphere that in turn control the flow of radiation in the atmosphere and hence determine the climate of the planet.

Even more than the land surface, the oceans control Earth's climate and the chemistry of its atmosphere. Their huge heat capacity regulates temperatures. Oceans also absorb carbon dioxide from the air, and their marine life absorbs and emits important components of Earth's chemical cycles. Satellite observations—of water colors due to suspended organisms (plankton), of ocean surface temperatures and of the winds that drive ocean currents—are critical to understanding our planet's evolution.

The atmosphere ties all of the processes together, providing pathways through which living systems can interact and exchange matter and energy. Some atmospheric patterns and processes are now readily observed from space; others, such as detailed wind fields and precipitation patterns, will become measurable in the *Eos* era.

Building Models of Earth

For some subsystems of Earth our understanding is good enough to identify critical processes and encapsulate them in computer models. However, for most cases we need to develop equations for computer models that are more comprehensive and more sensitive to the complexity of Earth's patterns and processes.

In all cases our knowledge is scale dependent. With vegetation and soil physics, we understand the processes better on the molecular scale than on the scale of a forest or savana. With the atmosphere and ocean, we know well the equations that describe the gross features and motions, but optimum descriptions of turbulence, air-sea interactions and eddy motions still elude us. In the geological realm, we have theories of mantle convection and crustal motion but are not confident that we can describe the evolution of the land surface quantitatively.

Transformation of patterns into knowledge of processes and their interactions is the crucial step. That done, we can feed data into computer models that will simulate the evolution of the Earth system and perhaps foretell its future. That is the hope and the goal of *Eos*.

Legacy of *Eos*

Future generations, rather than we, will benefit from the knowledge gained through *Eos* and the Mission to Planet Earth. They will know vastly more about the threats of ozone depletion and greenhouse warming, and more about their planet. They will have a science of the entire Earth system. As benefits of space observations and technology, there will be greatly improved weather prediction with benefits to public safety as well as agriculture and commerce. Increased knowledge about the dynamics of the crust will improve understanding of earthquakes and how to predict them. Future generations will manage energy resources better and will manage, or at least be more aware of, the vagaries of the global water cycle. They will be more sensitive to the potential for human activities in one area having important consequences for all humans, for the entire planet.

If we attain the goals of *Eos* and Mission to Planet Earth, then future generations will know that we, as the 20th century came to a close, accepted our responsibility for the future.

"First in space will mean first on Earth. And America intends to stay No. 1."

Ensuring America's Technological Superiority Should Be the Main Goal

National Space Council

Many nations, including those comprising the European Space Agency, seek to profit from space exploration. In the following viewpoint, the National Space Council argues that to compete with these nations, the United States must strive to become the world's strongest spacefaring country. The council asserts that leadership in outer space will reward America's industry and economy with scientific and technological innovations that can help to guarantee America's superiority on earth. The council, part of the executive branch of the federal government, coordinates U.S. space policies and strategies. It comprises many members of the president's cabinet.

As you read, consider the following questions:

1. According to the council, how can space exploration satisfy America's future energy needs?
2. Why do the authors believe that outer space is important to national security?
3. How can space exploration lead to medical breakthroughs, according to the council?

Adapted from the National Space Council's *1990 Report to the President*.

Since signing the Executive Order that established the National Space Council, President Bush has made clear his resolve that this nation will lead the world in space. His landmark speech on July 20, 1989, the 20th anniversary of the Apollo Moon landing, established America's goals in space exploration: ". . . a long-range continuing commitment . . . first for the coming decade—Space Station Freedom—our critical next step in all our space endeavors. And next, for the new century—back to the Moon. Back to the future. And this time, back to stay. And then—a journey into tomorrow—a manned mission to Mars. . . ."

The President reaffirmed his resolve in his speech at the University of Tennessee on February 2, 1990: ". . . first in space will mean first on Earth. And America intends to stay No. 1. . . . Our goal: To place Americans on Mars—and do it within the working lifetimes of the scientists and engineers who will be recruited for the effort today. . . ." A subsequent speech at Texas A&I University on May 11, 1990 put a firm date—2019—on his goal: "I believe that before Apollo celebrates the 50th anniversary of its landing on the Moon the American flag should be placed on Mars."

U.S. Space Strategy

The Vice President, too, has demonstrated publicly his strong support for the President's space objectives. He enunciated U.S. National Space Strategy in a major address to the American Astronomical Society on January 10, 1990: "First, we intend to develop our space launch capability and its related infrastructure as a national resource. . . . Our second goal is to open the frontiers of space. This includes manned and unmanned programs. . . . [Third,] the National Space Council is committed to intensifying our use of space to deal with problems on Earth. . . . [Fourth,] we believe the exploration of space will enhance our economic well-being and our overall national competitiveness . . . and the final element of our strategy, of course, is ensuring that our space program contributes to our nation's security. . . ."

Having defined the Space Council's planning process for implementing the President's goals in space, the Vice President laid before the American people his rationale for a strong and comprehensive civil space program. At the U.S. Space Foundation's Sixth National Space Symposium in Colorado Springs on April 10, 1990, he said ". . . in the next century space may be key to allowing us to satisfy our energy needs from space without damaging the environment; providing us with increased access to rare and essential metals and minerals; and allowing us to develop new information services which could further the revolution in communications which has already begun. And of course, in the next century, research in space could lead to new medicines or medical treatments of incalculable benefit to mankind.. . . ."

U.S. Space Effort and National Strength

We have regained the initiative in the exploration of outer space . . . making it clear to all that the United States of America has no intention of finishing second in space.

This effort is expensive, but it pays for its own way, for freedom and for America. . . . There is no longer any doubt about the strength and skill of American science, American industry, American education, and the American free enterprise system. In short, our national space effort represents a great gain in, and great resource of, our national strength.

John F. Kennedy, quoted in *Twenty-First Century*, May/June, 1989.

Three weeks later, at the Annual Meeting of the American Institute of Aeronautics and Astronautics in Washington, D.C. on May 1, 1990, the Vice President emphasized, "Our future competitiveness will depend on advancing technology . . . on educating our young people for excellence in math and science. The space program is a sound investment in ensuring that these key aspects of American competitiveness are there when we need them.". . .

Key Strategy Elements

The Space Council's approach for implementing U.S. national space policy divides all space activities into five areas, each of which may encompass civil, national security, and commercial activities conducted by NASA, DOD, DOE, DOT, other government agencies, or the private sector.

The space program serves multiple objectives: preserving the nation's security; creating economic opportunity; developing new and better technologies; attracting good students to engineering, math, and science; and exploring space for the benefit of mankind. The Council's approach is designed to achieve these objectives as an integrated national effort cutting across traditional lines of civil, national security, and commercial programs. The five key elements of U.S. National Space Strategy are:

1. To develop U.S. space launch capability—our transportation to and from space—as a national resource: the space transportation infrastructure will be to the 21st century what the great highway and dam projects were to the 20th. We will ensure that this infrastructure provides assured access to space, sufficient to achieve all U.S. space goals.

49

2. To open the frontiers of space through both manned and unmanned exploration: we will build on the successes of Viking and Voyager and proceed to comprehensively explore the solar system with Magellan, Hubble, Ulysses, and other ambitious unmanned programs. The President's call to complete Space Station Freedom, return to the Moon to stay, and the journey to Mars has finally given a much needed focus to our manned efforts. New ideas will be synthesized into varied approaches to undertake these premier space flight missions of the future.

Security and Well-Being

3. To intensify our use of space in solving problems here on Earth: we already use space systems to verify arms control treaties and to provide our defense forces with warning, communications, navigation, meteorology, and other functions vital to our national security. Satellite communication networks link peoples around the globe and contribute to the increasingly successful fight against repression and totalitarianism. Remote sensing from space contributes to a variety of land and ocean use applications and helps us understand, and potentially mitigate, the process of global climatic change.

4. To foster our economic well-being: we will capitalize on the unique environment of space to investigate and produce new materials and medicines and develop clean and abundant energy for all. The resulting private investment will create jobs; boost the economy; and strengthen our science, engineering, and industrial base. Along the way, new commercial space markets will be created and existing industries will become stronger and more competitive in the world marketplace.

5. To ensure the freedom of space for exploration and development: there are currently numerous spacefaring nations, with many others on the way. Space will become to the future what oceans have always been—highways to discovery and commerce. But the sea lanes must be open to be usable, and as we know from past conflicts, they are subject to disruption. Thus, we must ensure the freedom to use space for exploration, development, and security for ourselves and all nations. . . .

The Space Exploration Initiative

The Space Exploration Initiative [SEI] was a major activity of the Space Council during its first year. When President Bush announced his long-range goals for human exploration of the Moon and Mars on July 20, 1989, he asked the Space Council to develop a strategy for achieving these goals.

Various detailed programs for permanent settlement of the Moon and the human exploration of Mars have been proposed for over 20 years as logical extensions of the capabilities we developed for Apollo and subsequent Earth-orbit operations.

The President's July 20, 1989 announcement firmly established the nation's long-range goals in the human exploration of space: to proceed from Space Station Freedom to a permanent lunar presence in the next century, followed by a mission to Mars.

The Space Exploration Initiative is the ultimate investment in America's future. By responding to the human imperative to explore, we will reap benefits for ourselves and future generations akin to those of the voyages by Columbus and Magellan. We will increase our storehouse of knowledge about the planets, including our own, and about the nature of life itself. We will develop new technologies, many of which will have applications that will improve our lives on Earth. We will stimulate science and engineering education in this country by inspiring and motivating our young people. And we will be setting the stage for eventual permanent human habitats on other planets. Moreover, the Space Exploration Initiative will improve our competitive technological position in the world while enhancing our national pride and international prestige. But most importantly, the technological capabilities we develop, the new resources we discover, and the new industries we find in pursuit of these ambitious space-exploration goals will power American economic preeminence throughout the 21st century.

Recommendations for Action

The Space Council was charged by the President to define an approach by which his space exploration goals could be best achieved, including an assessment of the possibilities for international cooperation.

The Council received suggestions for implementing the initiative from NASA, the Department of Energy, and industry firms. The Council also benefitted from a review of these ideas by the National Research Council. All concluded that the President's goals were achievable, but the suggested approaches varied widely. The Council decided that so ambitious a program would require a systematic search for innovative concepts and new technologies having the potential to reduce costs, accelerate schedules, and reduce risk.

The Space Council therefore recommended to the President that such a search be conducted and that the Initiative first focus on technology development. The Council also concluded that at least several years should be devoted to defining two or more significantly different program architectures and developing and demonstrating technologies broadly applicable to space exploration. The President accepted the Council's recommendation, and on February 16, 1990 it issued a policy directive to that effect.

The decision also stated that the initiative be led by NASA and include a balanced program of robotic and manned exploration missions. To take maximum advantage of existing capabilities, however, the technological and systems expertise of other relevant agencies should be tapped. Therefore, the Departments of Defense and Energy will play significant roles in technology development and concept definition and will continue to work with NASA to develop the SEI.

Maintaining World Leadership

Ours is a rapidly changing world. To remain competitive and maintain world leadership in the 21st century, America will need the best trained and educated work force, the most advanced technology and the strongest leadership. We now have goals that challenge our abilities far beyond what we've experienced before.

The Space Exploration Initiative is a vision for the 21st century. It is a vision of America reaching beyond itself, and onward, beyond the very bounds of this planet to an entirely new world. On the way there, we will reap the real, tangible benefits of space exploration.

Thomas P. Stafford, *America at the Threshold*, 1991.

To define our technology development and architecture development programs, the Space Council has chartered an Exploration Outreach and Synthesis activity. Throughout the summer of 1990 a number of government agencies, professional technical organizations, federally contracted research centers, and private citizens developed ideas on the technologies and approaches which could enable us to accomplish our exploration goals faster, cheaper, and better. . . .

America's Future at Stake

The National Space Council is evaluating alternative strategies for using space to benefit mankind. Council activities have focused on validating, expanding, and articulating the National Space Strategy and extending its guidance to specific opportunities throughout the space community.

Our efforts are guided by several specific principles. First, the United States plans to develop and pursue its opportunities in space. Space exploration and the application of space technologies is as much an imperative for the continued development of the nation as was exploration of the continent by our forefathers. America's future is inevitably and irrevocably linked to our efforts in space. This reality underlies the Council's sense of urgency in approaching its tasks.

Second, one of the greatest strengths of this nation is its ability to meld the efforts of its technological, industrial, academic, and governmental institutions toward a common cause. The Council's policies and plans for space capitalize on this strength by seeking to map a course that harnesses the innovative, creative and analytic prowess of all American institutions. Consequently, the Council treats each goal, each objective, and each initiative as a joint undertaking.

Finally, the Council's actions recognize that what is needed is not leadership in space per se, but leadership in using space to address important human concerns. Space offers unlimited potential for improvement in almost every area of human endeavor; such as in medicine, where microgravity may offer new and undreamed of pharmaceuticals and cures; in industry, where productivity may be increased and made more affordable; and in national security, where space capabilities allow us to verify arms treaty provisions and counter threats to the nation's well-being. Space also offers tremendous potential for new sources of needed materials and clean, unlimited energy.

Challenges in Space and on Earth

The excitement and challenge of the space program, both manned and robotic, can be a powerful motivator for young people to enter science, engineering, and technology fields. This was clearly shown during the Apollo moon program, when advanced degrees in these fields rose dramatically in response to the program's investments. The Space Council plans to continue to emphasize education as an integral part of the space program.

The President and Vice President have given America a clear vision of a bright, prosperous future. It is a vision based on the unlimited potential of space to benefit mankind; but one that can be realized only if all Americans commit themselves to U.S. leadership of a global campaign to explore space, understand and appreciate it, and harness it in service to mankind.

Today, America is faced with tremendous, all-pervasive challenges—in medicine, in energy, in industrial competitiveness, in national security, in the environment, and elsewhere. How well we meet these challenges will determine how we and all citizens of the world live in the future. The real questions that confront us are whether we appreciate their urgency, whether we understand the potential of space in meeting them, and, understanding that potential, whether we as a nation are willing to make the major commitments necessary to engage in the exploration of space with the dedication and seriousness that these challenges demand.

Fortunately, America and the world now have new opportunities to consider and act on these challenges. The relaxation in world tensions and the resulting spirit of cooperation permeating all of Europe, extending even to the Soviet Union, offer unprecedented opportunities for realigning our nation's scientific and technical resources toward space and for achieving true cooperation among nations. Surely now is the time to capitalize on these opportunities.

A Solid Foundation in Space

It is with these thoughts in mind that the National Space Council has undertaken its efforts. While the Council recognizes the urgent need for increased commitment in space, it also appreciates the unrelenting need to put our national and international space activities on a sound footing once and for all. Only such a foundation can effect an efficient transition from a space effort geared toward research to one that applies the potential of space to solve mankind's problems and assure our nation's future.

To begin the process of establishing that sound footing, the Council implemented its charter by setting up a plan that identified four phases: setting broad goals and objectives for America in space, establishing strategies for achieving those goals and objectives, monitoring the implementation of those strategies, and resolving specific program or policy issues.

To undertake these functions we articulated America's space strategy as a set of critical elements for attaining the benefits space offers. We have defined these major elements and outlined the various investigations we have conducted or are in the process of undertaking. The sole purpose of all our activities is to determine where we are and where we must go to take maximum advantage of the opportunities offered by space exploration and exploitation.

Distinguishing Between Fact and Opinion

This activity is designed to help develop the basic reading and thinking skill of distinguishing between fact and opinion. Consider the following statement: "The *Viking* lander failed to find signs of life on Mars." This is a factual statement because it could be checked by referring to an encyclopedia. But the statement "The *Viking* mission was a failure" is an opinion. Many scientists believe *Viking* accumulated important data about Mars. Simply because the probe did not discover traces of life there does not necessarily mean the mission was a failure.

When investigating controversial issues it is important that one be able to distinguish between statements of fact and statements of opinion. It is also important to recognize that not all statements of fact are true. They may appear to be true, but some are based on inaccurate or false information. For this activity, however, we are concerned with understanding the difference between those statements that appear to be factual and those that appear to be based primarily on opinion.

Most of the following statements are taken from the viewpoints in this chapter. Consider each statement carefully. *Mark O for any statement you believe is an opinion or interpretation of facts. Mark F for any statement you believe is a fact. Mark I for any statement you believe is impossible to judge.*

If you are doing this activity as a member of a class or group, compare your answers with those of other class or group members. Be able to defend your answers. You may discover that others come to different conclusions than you do. Listening to the reasons others present for their answers may give you valuable insights into distinguishing between fact and opinion.

> O = *opinion*
> F = *fact*
> I = *impossible to judge*

1. Life exists in other galaxies.

2. Gravitational forces on the Moon match those on Mars.

3. Satellites increase humans' knowledge of earth's environment.

4. People will thrive in outer space.

5. NASA actively searches for extraterrestrial intelligence.

6. Scientific research yields more benefits than space exploration.

7. Studying earth's environment is more important than searching for extraterrestrial life.

8. Space technology generates useful products and services.

9. Water is present on Mars.

10. The greenhouse effect harms life on earth.

11. Long-duration space flight causes physiological problems for humans.

12. The Moon presents fewer problems to colonization than Mars does.

13. Scientists search for extraterrestrial intelligence by listening for radio signals that originate in outer space.

14. Few Americans will actually colonize outer space.

15. Planetary mining is an impractical goal.

16. One of the early uses of satellites was to study agricultural conditions on earth.

17. Apollo astronauts brought lunar rock samples back to earth.

18. Experiments in space do not require the presence of astronauts.

19. NASA is wasting its time and money searching for extraterrestrial intelligence.

20. Many private corporations have expressed interest in developing outer space resources.

21. Scientific research is NASA's most important priority.

22. In 1992, NASA predicted that a hole in earth's upper atmosphere ozone layer would soon develop over parts of the Northern Hemisphere.

Periodical Bibliography

The following articles have been selected to supplement the diverse views presented in this chapter.

Isabel S. Abrams — "Zoom with a View: Mission to Planet Earth," *Current Health 2*, April 1991.

Shawn Carlson — "SETI on Earth," *The Humanist*, September/October 1991.

Leonard David — "Moon Bases Debated," *Ad Astra*, June 1991.

Marsha Freeman — "The Science Needed to Understand Climate," *Twenty-First Century*, Winter 1990.

Marsha Freeman — "Sending Man into Space Is the Key to Economic Growth," *Twenty-First Century*, Summer 1991.

Peter Garrison — "Harvested Moon," *Omni*, April 1990.

Bill Green — "Earth in Harm's Way: What Can We Do to Help?" *USA Today*, May 1990.

John Halford — "Are We Alone?" *The Plain Truth*, February 1991. Available from PO Box 111, Pasadena, CA 91123.

Mickey Lemle — "New Horizons," *New Age*, March/April 1990.

Gerard K. O'Neill — "Living in Space," *The World & I*, January 1992.

Kumar Ramohalli — "Mining the Air: Resources of Other Worlds May Reduce Mission Costs," *The Planetary Report*, January/February 1991. Available from The Planetary Society, 65 N. Carolina Ave., Pasadena, CA 91106.

Skeptical Inquirer — "Searching for Extraterrestrial Intelligence: More Viewpoints," Fall 1991.

Technology Review — "Rethinking Space: An Interview with Norman R. Augustine," August/September 1991.

John M. Williams — "The Starhunters," *Final Frontier*, November/December 1991. Available from PO Box 534, Mt. Morris, IL 61054-7852.

William F. Wu — "Taking Liberties in Space," *Ad Astra*, November 1991.

Which Space Programs Should the U.S. Pursue?

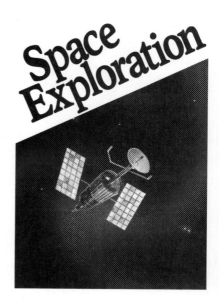

Chapter Preface

December 11, 1972, marked the last landing of American astronauts on the moon. Since then, human spaceflight has been primarily limited to space shuttle research and commercial satellite missions. In 1990, however, George Bush called for expanding human spaceflight as a goal of U.S. space exploration. Bush demanded that NASA work toward having astronauts return to the moon and land on Mars by the year 2019. Current NASA projects are aimed at ensuring that this goal is met.

Bush's announcement generated a great deal of controversy. Space exploration enthusiasts applaud Bush's efforts. They assert that although it is costly to sustain human life in space, these costs are justified because there are no substitutes for astronauts' skills. Jerry Grey, policy director of the American Institute of Aeronautics and Astronautics, writes, "At a certain level of mission complexity it becomes cheaper to invest in the life-support system and safety assurance needed to send humans to do the job."

However, opponents of manned expeditions believe that unmanned spacecraft are superior modes of exploration. They cite the success of Voyager expeditions that have accumulated valuable data of the far planets for more than a decade. They also point out the potential of using robots to explore the moon and Mars. Robots would be immune to the harsh environment of space, and there would be no risk to human lives if an accident occurred, they contend. As physicist and science writer Shawn Carlson writes about a Mars mission, "A robot mission would safeguard more than just astronauts. If people died in space, the whole Mars program would likely die with them."

For many years, American astronauts have successfully explored space and paved the way for others to follow. But the death of astronauts aboard the space shuttle *Challenger* proved that such accomplishments can come at a high price. The following chapter presents arguments concerning the role of astronauts in future space missions and what types of space missions the United States should pursue.

"Human exploration of space, leading to human settlement beyond the earth, is . . . an investment in the future of humanity."

The U.S. Must Pursue Manned Space Exploration

National Space Society

In the following viewpoint, the National Space Society (NSS) argues that humans must voyage into space because exploration of unknown regions has historically created new opportunities for humanity. The society maintains that U.S. funding of manned space exploration generates scientific and technological advancements. The NSS also states that expansion into space guarantees the survival of the human species, which could become extinct from nuclear war or other catastrophes. The NSS is a space exploration advocacy group located in Washington, D.C.

As you read, consider the following questions:

1. How can robots aid manned space exploration, according to the NSS?
2. Explain the authors' belief that manned space exploration is part of the evolution of life.
3. In the society's opinion, how is manned space exploration comparable to the European settlement of the Americas?

Adapted from *Response of the National Space Society to the Report of the Advisory Committee on the Future of the U.S. Space Program*, an NSS Position Paper (January 1991); principal author, Assoc. Prof. Glenn H. Reynolds, University of Tennessee Law School, NSS Legislative Committee Chair. Reprinted with permission.

The National Space Society, with over 30,000 members and over 120 chapters across America, is the nation's largest and oldest pro-space group. Our goal is the creation of a spacefaring civilization, one in which space travel will be affordable, routine, and commonplace, and in which human settlement exists in a variety of locations beyond the earth. This is a goal that has explicitly been endorsed by the Congress, in the Space Settlements Act of 1988, and by the Executive, in President Bush's speech of July 20, 1989.

In light of our goals, NSS naturally has a strong interest in the progress of America's civil space program. For this reason, we applaud the creation of the Advisory Committee on the Future of the United States Space Program, and we wish to express our gratitude to the Committee for the effort and time that it has devoted to this important task. And NSS is generally quite pleased with many of the Committee's recommendations. We are particularly happy that many specific recommendations made by NSS in its testimony before the Committee—such as the importance of establishing a lunar base prior to any human mission to Mars, the need for an aggressive space technology development program, the importance of fostering the growth of the commercial space sector, and the priority of life sciences research for space station *Freedom*—were adopted by the Committee. . . .

Robotic Exploration

In general, NSS is disappointed at the Committee's emphasis on robotic missions over human exploration. Although the report of the Committee contains excellent language on the importance of human exploration and colonization throughout outer space, the Committee's specific recommendations do not always reflect these views. Although it is not entirely clear, the report appears to favor the "Mission to Planet Earth" Earth Observing System as a top priority, to be followed by robotic missions of exploration to other planets, and—on what the Committee calls a "go as you pay" basis—missions of human exploration. . . .

It is not that robotic exploration is unimportant, or that NSS does not support it—in fact, NSS was the *only* pro-space group to enter the legal battle in support of the *Ulysses* launch, notwithstanding that other groups, such as the Planetary society, devote more of their energies to talking about the importance of robotic missions. However, NSS believes that the primary importance of robotic exploration is a precursor to human exploration. Just as specialized exploration missions—like those of Lewis & Clark or Zebulon Pike—preceded human settlement of America's west, so robotic missions should have the primary purpose of gathering data that will facilitate later human exploration and settlement.

Some critics of human exploration seem to believe that the

only legitimate purpose for robotic missions is the gathering of pure scientific data, unconnected with any later human missions. While NSS certainly has no quarrel with the gathering of scientific data (and we note that "abstract knowledge" always turns out to have concrete uses in the end), we would not consider a world in which everything was known about the Solar System, but in which no human left the surface of the earth, to be an acceptable one. Such a world might be acceptable to some (though certainly not all) planetary scientists, but we must note that their opinions are hardly objective, given the tangible rewards—such as tenure, research money, and professional standing—that flow from the unmanned missions they champion. The Committee was right to reject such an approach.

The Right Balance

If America is lulled into believing that unmanned space exploration missions are a reasonable and budget-conscious substitute for a manned space station, a space transportation system and an eventual moon base, then we are effectively cutting ourselves off from the primary frontier of the 21st century. Our space science program was not intended to replace our manned space program. It was designed to be a companion to manned space efforts. Together the two form the exploration and the development balance needed for continuous space expansion.

Robert A. Roe, *The American Legion Magazine*, July 1990.

NSS recommends a broader view. While learning about our solar system is worthy in itself, the greatest returns to humanity as a whole will come only if that knowledge is put to use in a more concrete way: to further the goal of human settlement beyond the earth. Only in this context does a truly major commitment to unmanned planetary exploration make sense. Human exploration and robotic exploration must go hand-in-hand, notwithstanding the all-too-popular view that they should be thought of in either/or terms. In fact, "either/or" thinking on this topic is unlikely to help even the cause of robotic exploration: if human exploration and expansion are given a low priority, it then becomes more difficult to make a strong case for robotic missions either, given that other competitors for research dollars—biomedical research, for example—might then arguably have a stronger claim to improving the human condition than would the abstract knowledge generated by robotic planetary missions. The result would be a disaster for the space community as a whole, and for humanity. For this reason, we oppose treating robotic exploration as somehow separate from, and independent of, human exploration and settlement.

Unfortunately, the Committee's report at least appears to encourage such a view, by treating human exploration—and only human exploration—as a part of the program to be funded only to the extent that surplus money is available. NSS is somewhat doubtful that this is what the Committee members really meant by their language concerning a "go as you pay" approach, but this is certainly the way that that language has been interpreted, and will continue to be interpreted unless the Committee members make clear that this was not what they meant.

At any rate, the reliance on a "go as you pay" approach is a bad idea. Human exploration of space, leading to human settlement beyond the earth, is not a consumption item, but an investment in the future of humanity. But, like most investment programs, it is less fun than consumption in the short term. Thus, just as it is poor strategy for individuals to plan their savings programs based on putting away whatever "extra" cash they may happen to have on hand in the future, it is poor strategy to plan on funding human exploration programs by relying on "extra" cash in the federal budget. As the Committee itself notes, *predictable, long-term funding is absolutely essential* to the success of space projects. Opportunistic funding approaches, as suggested by the "go as you pay" terminology, are unlikely to generate such funding.

Furthermore, as the Committee also recognizes, NASA has done its best where its work was in pursuance of a major, articulated national goal that was explicitly given a high priority. Again, a "go as you pay" approach seems inconsistent with this observation, and NSS is very concerned that "go as you pay" will be translated in practice to "stay home," with disastrous consequences for America and for humanity. Furthermore, in the absence of clear goals, explicitly provided for, we are doubtful that technology development and robotic exploration plans will have enough direction for optimal performance. Saying, "we will go to Mars someday, if we can afford it," simply isn't enough to provide much policy direction. . . .

Humanity in Space

The National Space Society's vision is of a spacefaring civilization. With humanity residing throughout the solar system, we will no longer be at risk of extinction because of a single event, whether it is a nuclear war or a cometary impact like that believed to have wiped out the dinosaurs. With access to cheap solar energy and the material wealth of the solar system, many problems of scarcity and environmental degradation will be solved. And, during that expansion, we will be carrying on the frontier spirit that made America great.

In the introduction to his excellent history of the Space Age, historian Walter McDougall compares humankind's emergence into outer space to the emergence of the first lungfish from the oceans. Though dramatic, this comparison is not overdrawn. The emergence of humanity into space represents a real qualitative change in human existence. Over the next century or more, if affairs on earth are not too badly mismanaged, humanity will spread to many locations outside the earth, carrying parts of the earth's biosphere with it.

An Inspiring Initiative

Human exploration of the solar system is a challenging and inspiring initiative that can promote global understanding, peaceful cooperation, scientific progress, technology development, and educational excellence. This massive effort will clearly include opportunities for participation by many countries. It will provide a focus for space science and technology programs in all of the spacefaring nations and enhance new initiatives in emerging nations.

Association of Space Explorers, *The SPACExplorer*, November 1991.

In fact, some writers have suggested that this is humanity's real role. If one believes in the *Gaia* hypothesis, which says that all life on earth can be viewed in some sense as one meta-organism, then perhaps humanity's role can be viewed as that of meta-gamete, carrying the seeds of life to new environments where life could not have evolved, and which it could not have reached in other ways. And, regardless of whether one accepts this analogy or regards it as so much science fiction, there is no question that space resources, in terms of both materials and energy, hold enormous promise for providing long-term relief to Earth's overstressed environmental and resource base.

The National Space Society believes that this expansion of humanity is the most important project at hand today. Furthermore, we believe that it is not only important in itself, but that it will create enormous benefits even for those of us who remain at home. Most people nowadays are familiar with the many economic and technical "spinoffs" from the space program. But these are not the only ones. In a society, like ours, that seems to lack direction and goals itself, the opening of a new frontier in space could have enormous—and positive—psychological impact. Such was the effect of Columbus' voyage, and the ensuing settlement of the Americas by Europe. As Samuel Eliot Morison recounts in his classic biography, *Admiral of the Ocean Sea:*

At the end of the year 1492 most men in Western Europe felt exceedingly gloomy about the future. Christian civilization appeared to be shrinking in area and dividing into hostile units as its sphere contracted. For over a century there had been no important advance in natural science, and registration in the Universities dwindled as the instruction they offered became increasingly jejune and lifeless. Institutions were decaying, well-meaning people were growing cynical or desperate, and many intelligent men, for want of something better to do, were endeavoring to escape the present through studying the pagan past. . . .

Yet even as the chroniclers of Nuremberg were correcting their proofs from Koberger's press, a Spanish caravel named *Nina* scudded before a winter gale into Lisbon, with news of a discovery that was to give old Europe another chance. In a few years we find the mental picture completely changed. Strong monarchs are stamping out privy conspiracy and rebellion; the Church, purged and chastened by the Protestant Reformation, puts her house in order; new ideas flare up throughout Italy, France, Germany, and the northern nations; faith in God revives and the human spirit is renewed. The change is complete and astounding. "A new envisagement of the world has begun, and men are no longer sighing after the imaginary golden age that lay in the distant past, but speculating as to the golden age that might possibly lie in the oncoming future."

Christopher Columbus belonged to an age that was past, but he became the sign and symbol of this new age of hope, glory, and accomplishment. His medieval faith impelled him to a modern solution: expansion.

This experience, it seems to us, is a good enough answer to those who say that we should solve our problems at home before we go into space. Where would we be today if the Europe of Columbus' time had felt the same way?

Human Exploration Must Begin

It is past time to start pursuing our objectives in a serious way. We have seen numerous studies on how to get into space, and what should be done, and inadequate progress on implementing them. The National Space Society has declared that the 1990s should be a "decade of doing." For this reason, we hope that the Committee's report will largely be adopted, but with a real sense of the ultimate reasons why the space program is important to begin with. We look toward solutions and approaches that will bode well for our future, on this world and the worlds yet to come. The time for study is past; the time for action is now.

"Manned spaceflight is a feel-good adventure the country can ill afford."

The U.S. Must Not Pursue Manned Space Exploration

Alex Roland

The cost of sustaining human life in space is far higher than that of unmanned space exploration. In the following viewpoint, Alex Roland argues that because of this high cost, the United States should not waste money by sending astronauts into space. Roland states that the United States can explore space and perform experiments much more efficiently with unmanned spacecraft. Roland, a former NASA historian, is a history professor at Duke University in Durham, North Carolina.

As you read, consider the following questions:

1. How do astronauts hamper space research and spaceflights, according to Roland?
2. In the author's opinion, how are satellites proof of the success of unmanned spacecraft?
3. Why does Roland believe that the United States has no need to prove its superiority in space?

Alex Roland, "The Case Against Manned Space Flight." Position Paper, February 1992. Reprinted with the author's permission.

Manned spaceflight is a feel-good adventure the country can ill afford. There are at least four reasons for cutting back dramatically on the flight of people into space.

First, manned spaceflight has turned out to be more expensive and more difficult than predicted. The Apollo program made it look as if it were easy and profitable. But Apollo cost $25 billion, which would be approximately $80 billion in today's dollars, and the mission turned out to be a dead-end. We got the international prestige we bargained for and we won the space race. But the public lost interest in the missions and the scientific payoff was less than we could have gotten by investing the same funds in unmanned activities.

Space Station and Shuttle Overspending

The story has been the same since. The space shuttle was supposed to cost $33 million per flight (in 1991 dollars); through fiscal year 1991 it has cost $1.5 billion per flight, a cost overrun of 5000%. The shuttle was supposed to open up space for greater exploitation by carrying payloads into orbit for $350 a pound. It has cost $32,000 a pound, essentially closing off space activity to all but communication satellites and highly subsidized government payloads. In addition to being the world's most expensive launch vehicle, the shuttle is also the most fragile and unreliable. It suffers repeated launch delays on the pad and is more vulnerable to weather than expendable launch vehicles. It has never carried the 65,000 pounds of cargo for which it was designed. It cannot land at any 10,000-foot runway. It cannot turn around in two weeks and be ready for another launch. The cost of making it safe enough for humans to fly is to make it more expensive than it is worth.

The space station already exhibits many of the same weaknesses, even before its construction begins. When President Reagan proposed the space station in 1984, he projected its cost at $8 billion. Eighteen months later the National Academy of Engineering estimated that it would really cost $32 billion. Since then the cost estimates have continued to rise, even while the station has been scaled back and its capabilities reduced. Like the shuttle, it now costs far more and can do far less than was initially predicted, and the worst is yet to come. The real cost overruns on the shuttle came in construction and operation. The same is likely to be true of the space station. And its only real mission, the only one that cannot be done more effectively and more economically by automated, unmanned spacecraft, is manned spaceflight as an end in itself. The primary purpose of this enterprise will be to try to solve the overwhelming problems of physiological debilitation that humans experience on long-duration spaceflights. For that dubious objective the civilian space program will be mortgaged to a white elephant in orbit for decades to come. As with the shuttle, the only thing more expensive than actually building it will be trying to operate it.

Unmanned Spacecraft

Second, most of the payoff in space has come not from manned vehicles but from unmanned. Virtually the entire military space program has been flown on unmanned vehicles. That unmanned program launched reconnaissance satellites which diminished the danger of the nuclear arms race by virtually eliminating the possibility of surprise attack. The program also erected a satellite communication system that gave commanders immediate access to forces and data around the world. It sustained a satellite navigation system that gave our ships pin-point navigation and even allowed our troops in the Gulf War to navigate across the pathless desert. All of these capabilities and more were emplaced without the necessity of people in orbit. The Air Force did not want the shuttle in the first place. It agreed to back shuttle development only on the coaxing of NASA; it reluctantly abandoned its expendable launch vehicles in favor of shuttle flights only at the direction of the White House. In the wake of the disastrous *Challenger* accident, which imperiled national security by cutting off access to space, the Air Force happily returned to primary reliance on its stable of expendable launch vehicles.

Expensive Manned Missions

As the ranking Republican member of the U.S. Senate Subcommittee on Science, Technology and Space, I have been a strong supporter of space research for national defense, technological spinoffs and employment benefits.

However, along with a number of my colleagues, I question whether America's continued pre-eminence in this field requires prohibitively expensive manned missions.

Larry Pressler, *The Washington Times*, August 14, 1990.

So too with space science, which has landed on Mars, visited all the planets in the solar system, and sent probes into deep space all without people on board. Indeed, as the scientists often emphasize, people on scientific spacecraft jar their instruments, interfere with delicate experiments, and limit the duration and itinerary of the spaceflights the scientists would like to conduct. Unmanned applications satellites have had similar success without people on board. They speed our communication around the globe. They take pictures of the earth for our daily weather reports. They monitor pollution, natural resources, and atmospheric conditions to assist in environmental research and control. They provide economic data on crop developments, irrigation, and oil exploration. And they provide all these services reliably and economically without astronauts. Any specific function that can be identified in space can be done more cheaply and more reliably by machines than by people.

Life Support Elevates Cost

Third, people are actually a constraint on most spaceflights, not an advantage. Once people are imposed upon a spaceflight, the mission of the flight changes. It is no longer scientific research or data gathering or satellite deployment. The primary mission becomes returning the people alive. All else must be subordinated to that goal. Money that might otherwise go into scientific or communications equipment gets diverted to life support systems. Launch rules change to provide greater safety for the crew, more safety than is necessary or practical with unmanned spacecraft. The duration and itinerary of the flight is constrained by the food and water on board; machines can fly near Mercury and beyond the outer planets, but people cannot. Precious weight on the spacecraft has to be given over to life-support, back-up systems, and radiation protection that would be unnecessary with machines.

The argument that people make up for the costs of their passage is simply specious. Unless the astronauts are going to use their senses of taste and smell, their role can be performed better by machines. Indeed, even the astronauts *in situ* use machines to do their experiments. Having the people in place provides only marginal advantage over having the machines run remotely by controllers on Earth. The advantage of an astronaut with a screwdriver who can respond to unexpected emergencies is far outweighed by the cost of putting him or her in place. It is far cheaper and more effective to simply build redundancy into our unmanned spacecraft so that if one system fails another will take its place. This is the technique which allowed us to put landers on Mars in the 1970s. In that case, we built two complete spacecraft and sent them both to Mars. Both got through, both deployed their landers, and both conducted experiments far beyond their predicted life expectancies. The astronauts on the moon did no better; indeed their exploration was more constrained. But those Apollo missions cost 25 times as much as the unmanned Mars missions.

Finally, the greatest practical payoff from manned spaceflight has diminished with the end of the Cold War. The real goal of the Apollo program was prestige, a demonstration that the

United States was technologically superior to the Soviet Union. That case was made in the 1960s. In the subsequent two decades, there was an attempt to argue that it had to be made over and over again. We had to have a shuttle because the Soviets were developing *Buran*. We had to have a space station because the Soviets had *Mir*. We had to go to Mars because the Soviets were going to Mars.

The Soviet Space Program

But the Cold War is over. The Soviet Union is no more. The *Buran* has flown only once and seems unlikely to fly again any time soon. The *Mir* space station is on the auctioneer's block; we could have a space station for a pittance if we chose. The former Soviet Union is backing away from manned spaceflight as an expensive and useless artifact of the Cold War. Other countries that were considering manned programs are rethinking them as well. Only the United States remains committed to a large manned spaceflight program. The result is that the United States now spends more on its civilian space program than all the rest of the world *combined*. And the major burden of this cost is manned spaceflight. This is an expensive and pointless extravagance that the United States no longer needs and can no longer afford.

One day humans will fly to Mars, perhaps beyond. Humans may even establish colonies in space. But there is no rush. The likelihood that we will blow up this planet has been greatly reduced. The possibility of a catastrophic collision with another heavenly body is infinitesimal. The need to race anyone to Mars has disappeared. It makes far more sense right now to put our house in order on Earth before we attempt to people the heavens. By the time we do that we may well have the technologies in hand to make manned spaceflight truly practical. Who knows, we may even by then have some reason other than circus for sending people into space.

"The Space Shuttle remains one of the world's most reliable launch systems."

The U.S. Should Continue the Space Shuttle Program

Stephen M. Cobaugh

Since 1981, NASA has relied on the space shuttle to launch astronauts and satellites into outer space. In the following viewpoint, Stephen M. Cobaugh argues that the United States should continue the shuttle program because it is one of the best launch systems ever built. Cobaugh contends that despite the *Challenger* accident, the space shuttle is more reliable than Europe's Ariane rocket and rivals NASA's successful Delta booster. He maintains that the shuttle, which has launched almost half of the materials that the United States has sent into outer space, will be NASA's primary transportation system well into the next century. Cobaugh is the editor of *Space Age Times*, a bimonthly space news magazine.

As you read, consider the following questions:

1. Why has the shuttle disappointed some members of Congress and the media?
2. In Cobaugh's opinion, how have the media unfairly criticized the shuttle?
3. According to the author, why will NASA depend heavily on the shuttle in the future?

Adapted from "Space Shuttle: The First Decade" by Stephen M. Cobaugh, *Space Age Times*, May/June 1991. Reprinted with permission.

71

Although the space age had been underway for two decades, nothing could have prepared the one million spectators that were about to witness an extraordinary moment in the history of exploration. A truly new era in space flight was about to unfold in the blue skies above the swampy marshlands of Cape Canaveral.

As dawn came on April 12, 1981, the bright clear morning was about to signal a dramatic change in the way the United States had placed astronauts into space. NASA, after a disappointing two-day delay, was about to debut a multi-billion-dollar gamble—the world's first reusable spacecraft. Nine years in the making, they formally called it the U.S. Space Transportation System [STS]. It was to become better known simply as the Space Shuttle.

Revolutionary Spacecraft

At Kennedy Space Center's [KSC] Launch Complex 39A, where a more traditional rocket had carried men to the Moon, sat a remarkable winged spaceplane with a barn silo-sized white fuel tank and twin pencil-like booster rockets. The entire system, which centered on a revolutionary concept never tried before, was not only visually striking but represented a tremendous technological gamble. Based on theory and computer analysis, STS should work. The time for tests was over.

Some 41,000 guests, including 2,700 news media, were poised on the grounds of KSC at prime "bird watching" sites, while three-quarters of a million tourists lined the highways and causeways of the Space Coast. Hundreds of millions more waited to witness the historic event on television.

To reach this point, facilities at KSC were reshaped from many existing Apollo structures and buildings. In fact, it had taken many years to transition the workforce and buildings to support the new Space Shuttle system. Engineers at the center reflecting on these changes believe it was easier to build the original structures than to adapt them from the sleek Saturn V vehicle to the more complex, stubby configuration that replaced it.

The big moment came at 7 a.m. EST as Space Shuttle Columbia's three main engines and boosters roared to life, powering two astronauts toward Earth orbit. The column of orange flame and billowing smoke that trailed afterwards signaled a profound shift in how we view space transportation. . . .

Shuttle Setbacks

The afterglow from Columbia's milestone flight had long vanished by the time the Challenger accident brought the program to an abrupt halt in 1986. In the final analysis, the Space Shuttle has become both a symbol of American ingenuity and the fragility of complex machines. As such, it has attracted opinions as diverse as its payloads during the past decade and, at times, as emotional as the current debate over abortion rights.

The visible public nature of the Space Shuttle missions has been something of a mixed blessing for a highly successful program frequently punctuated by launch slips, hardware failures, leaks, and accidents by ground technicians. STS has delivered almost half of all of the mass put into space by USA, yet there has been a tendency for the mass media to focus on problems rather than what NASA calls "the big picture."

Objectively, some of the criticism garnered by the Shuttle has been self-inflicted. Early in the program, for example, NASA sold the system based on overly optimistic economics and a very ambitious manifest that has yet to be met. When these objectives could not be accomplished, critics in Congress and the media were quick to jump on the bandwagon.

Ambitious Hopes

For its part, NASA points out that the vehicle it ended up with at the end of a long series of compromises is not the Shuttle it had envisioned. The agency had wanted a Shuttle system with a greater payload capacity, more powerful engines, and liquid-fueled boosters. When it got none of these, space officials insisted its new spaceplane could still keep all of its promises. And that list of promises was extensive:

• Launch of 60 missions per year (the most missions ever flown in a calendar year was nine in 1985). That original projection was about one launch every week of the year.

• Baseline of seven Orbiters (only four have been on-line at one time).

• Construction of a third launch pad at Vandenberg Air Force Base. (The pad, completed at a cost of over $2 billion, was never used and abandoned due to safety concerns.)

• Construction of two Orbiter Processing Facilities. (The Orbiter Maintenance Facility has been converted to an OPF, one of the few bright spots on the list.)

• Adequate spare parts to support the program. (Spare parts have always been a problem and the Challenger investigation board cited this as a major concern. More spare parts have been funded in subsequent budgets.)

• Regular landings at Kennedy Space Center to lessen the turnaround time between launches. (Only six of the more than 40 Space Shuttle missions have landed in Florida. Routine landings were expected to occur beginning in 1991.

Despite these shortcomings, the Space Shuttle remains one of the world's most reliable launch systems, with a success-to-failure ratio of .974 (1 being perfect). In fact, the Shuttle rivals the Delta expendable booster as the best launch system ever built.

The European-constructed Ariane rocket, the only other major launch system designed in the 1970s, had five failures in the first 40 flights.

Percentage of All Manned Space Missions by Vehicle

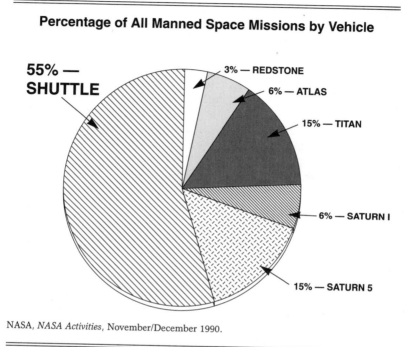

55% — SHUTTLE

3% — REDSTONE

6% — ATLAS

15% — TITAN

6% — SATURN I

15% — SATURN 5

NASA, *NASA Activities*, November/December 1990.

Today, excluding the Orbiter, STS has been responsible for launching 41% of all the mass the United States has put into space. In addition, it remains the only spacecraft capable of retrieving satellites from orbit and returning them to Earth. Even other spacefaring entities, like USSR and ESA [European Space Agency], have sought their own similar systems.

Henry Pohl, JSC's [Johnson Space Center] Director of Engineering, says in assessing the value of the Shuttle, comparisons to the much-touted Saturn V Moon rocket should be avoided. "The truth of the matter is if you go back and look at the number of people that worked on building those stages for every launch, and you translate that into today's labor rates, it'll cost you three times as much to launch one of those as it costs you to launch a Shuttle," he says. "And yet they say the Shuttle is too expensive."

This evaluation may be accurate, but the Space Shuttle will certainly have competition from another heavy lift expendable booster by the turn of the century. This new booster is expected to be part of NASA's new mixed fleet of vehicles recommended by advisory boards.

Even though many future projects will call for new projects like the National Aero-Space Plane and other advanced propulsion systems, the Space Shuttle will still continue to be the major focus of NASA's transportation needs well into the 21st century.

The next major task ahead for STS will be to support and resupply Space Station Freedom, with assembly set to begin in 1996. With the addition of Endeavour, Challenger's replacement vehicle, the fleet will be back to a full complement of four authorized Orbiters.

In the years ahead, the fleet will undergo constant modernization to reflect the latest technologies that state-of-the-art engineering can offer. These improvements, though, will not transform the Space Shuttle into the airplane-like operational "space truck" that NASA believed it could manage. Instead the fleet will remain a research and development vehicle, with all of its associated problems and shortcomings.

The Dream Is Alive

Aside from its setbacks, we can still marvel at what the Space Shuttle has accomplished since STS-1 and its unique contribution to science and technology. Its legacy has yet to be determined by history, but it's safe to say that it is playing an important role in cosmic exploration.

Astronaut John Young uttered an insignificant phrase during STS-1 which captured both the spirit and teamwork that went into making Columbia and her sister ships possible. Today, that phrase still says a lot about the state of affairs at NASA—"The Dream is Alive." It is, indeed.

"The shuttle is too costly, too complex, and too inflexible to support today's space access needs."

The U.S. Should Cancel the Space Shuttle Program

George A. Keyworth II and Bruce R. Abell

Critics of the space shuttle believe the United States should replace it with a new launch system. George A. Keyworth II and Bruce R. Abell, the authors of the following viewpoint, agree. Keyworth and Abell argue that the prohibitive costs of manned shuttle flights discourage its use and consequently hamper development of the space station and Mars programs which depend on the shuttle. The authors also assert that the infrequency of shuttle flights causes military and scientific projects to be delayed. They propose that the United States develop the National Aerospace Plane (NASP), a jetliner-like spacecraft that promises to fly more frequently and at less cost than the shuttle. Keyworth is director of research and Abell is a senior research fellow for the Hudson Institute, a public-policy research center in Indianapolis, Indiana.

As you read, consider the following questions:

1. Why do Keyworth and Abell oppose the use of a single launch system?
2. According to the authors, why are shuttle launches so infrequent?
3. Why do Keyworth and Abell believe that NASP is a safer alternative to the shuttle?

Adapted from "How to Make Space Launch Routine" by George A. Keyworth II and Bruce R. Abell, *Technology Review*, October 1990. Reprinted with permission from Technology Review, © 1990.

Since the last moon landing in 1973, the momentum of the space shuttle has dominated U.S. space launch programs. Here was a vehicle, we were told, that would make access to space routine. It would launch satellites, carry supplies to build a space station, and provide a zero-gravity platform for scientific research and, ultimately, industrial production. During the early 1980s, our national policy relied almost solely on the shuttle as our means of reaching earth orbit.

This policy was doomed from the start. The shuttle is too costly, too complex, and too inflexible to support today's space access needs. Moreover, overreliance on any single system leaves us extremely vulnerable in the event of an accident; after the Challenger tragedy, U.S. access to earth orbit virtually disappeared for almost three years.

After 20 years, it is no surprise that a system begins to look dated and inadequate. But addressing these realities head-on has often been awkward, even painful, because so much money and effort has been invested in the current launch systems and because there is no replacement system immediately on the horizon.

A Superior Launch Vehicle

But one research program now under way offers hope for precisely the kind of workaday access to space that shuttle proponents once envisioned: the National Aerospace Plane, or NASP. Unlike other launch vehicles that exist or are being developed, this aircraft would take off from a runway. It would then hurtle through the atmosphere at more than 20 times the speed of sound (Mach 20), deposit its payloads in low earth orbit, and finally descend and land on a runway.

NASP has been publicly perceived as primarily a hypersonic aircraft, intended for high-speed transport on earth, or as an exotic military reconnaissance or rapid deployment aircraft. (Conceived in the late 1970s by the Defense Advanced Research Projects Agency, NASP has been funded since 1985 through a joint NASA-Air Force program office at Wright-Patterson Air Force Base.) When the program first entered the public spotlight, much was made of the idea of an "Orient Express" that could carry passengers from New York to Tokyo in one or two hours.

But the technology's most immediate impact will be in space access. More than any other launch vehicle now being considered, NASP would provide low-cost and flexible access to space. Unfortunately, plans for post-shuttle space-launch systems have not yet seriously included vehicles using NASP technologies.

The United States faces a serious shortage of lift capacity. Right now the U.S. fleet is able to launch about a million pounds a year into low earth orbit. That does not even adequately cover the "official" launch demand compiled by the Air Force Space Command.

NASA'S CRACK FLEET OF SPECIALISTS AWAIT THE LANDING OF ANOTHER SUCCESSFUL SHUTTLE MISSION !

Ed Gamble. Reprinted with permission.

And demand will surely increase in coming years. The need for communications satellites will continue to grow as direct-broadcast television services are put into place around the world. Further demand will come from a proliferation of satellite-based navigation and position-locating systems; a new generation of earth-sensing programs, including the ambitious Mission to Planet Earth; and a backlog of planetary exploration projects. Moreover, military need for space access will increase, not decrease, as international tensions ease and surveillance supplants readiness as the basis for national security. When nations reduce defenses, they put a higher premium on intelligence—the old adage trust but verify.

Many of these missions could be served by unmanned launch vehicles. But if the United States seriously wants to build a space station or explore Mars—both proposed national goals—it will need a way to get people as well as payloads into orbit cheaply. Interplanetary manned exploration in particular will be unrealistic unless we reduce the cost of access to space.

But more importantly, there is a huge class of potential users

78

of earth orbit who cannot afford present launch systems. Each shuttle launch costs about $275 million, or $5,000 per pound of payload. Unmanned rockets are less expensive, but not dramatically; it costs about $150 million to put up a workhorse like the Titan, or about $3,000 per pound of payload. These costs form a high barrier to participation. Many more users will surface if prices drop to the $20 to $200 per pound range promised by NASP.

Providers of communications services other than for mass broadcast, for example, could take advantage of networks of satellites in low earth orbit (rather than in higher geosynchronous orbits). The idea of growing new kinds of materials in space—which so far has been essentially a stunt given the high cost of shuttle launches—could become economical.

Smaller Payload Trends

Shuttle launches are not only expensive, but also infrequent. Routine access means frequent launches and it means the ability to launch on relatively short notice. Yet after 40 years of space access with rockets, the United States is still a long way from that capability, even for unmanned systems.

Another element of routine access is the ability to carry a wide range of payload sizes. The space shuttle is designed to carry one or at most two large payloads. But this practice is anachronistic. Very few things that we want to put into space are the size of the Hubble space telescope. Most projected flight requirements would require a payload of no more than 25,000 pounds.

This shift toward smaller payloads is becoming particularly apparent in the growing popularity of small, cheap satellites. Conventional satellites, with their billion-dollar price tags, are the space-born equivalent of mainframe computers. Each multi-ton unit takes a decade (or longer) to build. Moreover, launch times must be reserved years in advance. Small satellites, typically weighing 50 to 1,000 pounds, are more like personal computers; they can be assembled quickly from inexpensive and accessible hardware and launched on short notice. And like the PC, as small satellites become more available, people will find unexpected new uses for them, further stimulating demand for launch services.

The move toward PC-equivalents in space is particularly important for scientists studying the earth. Missing from attempts to create realistic models of natural systems has been the ability to make large numbers of observations. We could learn more about the dynamics of global warming, for example, with 500 100-pound satellites than with one 50,000-pound satellite.

The shuttle's high cost and inflexibility have brought forth a number of proposals for alternative launch vehicles. But most of these would suffer from the same deficiencies that mark the shuttle. . . .

Unlike the shuttle or any of the other proposed alternatives, a NASP vehicle would offer frequent and low-cost access to low earth orbit without the complexities of a rocket launch.

NASP vehicles will use a jet engine to fly to hypersonic speeds and high altitudes. Equipped with massive air intakes, they should ideally reach Mach 22 (about four miles per second) in air-breathing flight. Rocket motors would accelerate the craft to Mach 24, the velocity needed to enter earth orbit.

Eliminating the need for rocket power from the ground to around 200,000 feet would reduce launch costs enormously. And unlike rockets, a NASP vehicle would not have to carry enormous tanks of liquid oxygen to use as the oxidizer for its hydrogen fuel. The craft will need to carry only enough on-board oxygen to fire the engines briefly for a final "kick" into orbit, to maneuver in space, and eventually to re-enter the atmosphere.

A NASP-derived vehicle would compare favorably with other space launch technologies. According to a May 1990 study by OTA [Office of Technology Assessment], the cost to launch an aerospace plane would run between $800,000 and $4.4 million per flight, for a payload of 20,000 to 30,000 pounds. That's far more economical than the shuttle, with its nominal launch cost of $275 million.

Airliner Similarities

The NASP's cost advantages stem primarily from its similarities to ordinary aircraft. Operating out of an airport and taking off from airport runways roughly the size used by commercial airliners, NASP vehicles could be launched frequently. According to plans developed by the NASP program office, a NASP vehicle could typically take off 24 to 36 hours after landing. Where rockets often sit in hangars and on the launching pad for weeks or months while the payload is loaded and all systems checked, NASP payloads and fuel could be loaded just a few hours before takeoff. A NASP vehicle will require inspection similar to that given an airplane—not the intense scrutiny required by a rocket. (If NASP suffers a mishap after takeoff, it can simply turn around and land again; the shuttle and other rockets risk total destruction.) And NASP will not be much more sensitive to weather conditions than airliners are. By contrast, the shuttle often waits days for acceptable launch conditions. Finally, while safety considerations usually force rockets to be launched over oceans, NASP could fly over land. This flexibility gives NASP the advantage of access to virtually all low-earth orbits.

NASP relies on two critical technologies: an engine that can achieve hypersonic flight with little or no rocket assist, and structural materials that withstand high stress and high temperatures. Recent developments in both areas are encouraging.

When NASP was first proposed as a national program in 1985, there was widespread skepticism that high-performance materials could be developed to hold up to the extreme operating conditions. This lack of suitable materials has essentially been solved.

"If we have a spiritual need for a new frontier—and I believe we do—Mars is it."

The U.S. Should Explore Mars

Michael Collins

Michael Collins was an astronaut on *Apollo 11*, the first mission to land astronauts on the moon. In the following viewpoint, Collins argues that the benefits reaped from the exploration and colonization of Mars warrant a U.S. expedition. Collins believes that the technological innovations derived from a Mars flight program would be a boon to the U.S. economy, just as spin-offs from the Apollo program contribute to industry today. He also maintains that people on earth could learn valuable lessons from a Mars colony devoid of disease, murder, war, and other problems. Collins, who lives in North Carolina and Washington, D.C., is an aviation and space consultant.

As you read, consider the following questions:

1. Why do humans have a strong desire for exploration, according to Collins?
2. According to the author, how can a Mars expedition provide information on the history of the solar system?
3. In Collins's opinion, why would Mars colonizers need to strive to work together?

If Mars is to become NASA's long-term goal, the reasons must be clearly understood and laid out in much greater detail than was the case for the Moon argued by John F. Kennedy. . . .

To me, the *why* of a Mars mission is rooted in the history of our planet and of this nation. I don't know whether exploration is in our genes, although I certainly think curiosity is, and the two are closely linked. At any rate, whether it is an inherited or acquired characteristic, most people have always gone wherever they could go. Some tribes have not, being content to live out their lives sealed off in remote valleys or hidden in rain forests, but even most primitive civilizations—Polynesians in their canoes, nomads on their camels, bushmen on foot across the Kalahari—have generally been wanderers. Certainly our European forebears, especially the Portuguese, Spanish, Italians, and British, were obsessed with reaching the farthest corners, if a globe can be considered to have corners. Many of their voyages were similar in duration to a Mars trip, and early explorers such as Ferdinand Magellan endured hardships that I hope will never be replicated in space. The great explorers were willing to put up with such long-term privation for a variety of reasons. Patriotism, ego, religion, inquisitiveness, greed—all played a part. When asked what he hoped to find in India, Vasco da Gama replied, "Christians and spices." There will be no Christians waiting on Mars and probably no spices, but eliminating religion and commercial return (at least for the first few landings) is not necessarily bad. The urge to go, to see, to touch, to smell, to learn—that is the essence of it, not to mention the exhilarating possibility of encountering something totally unexpected. . . .

Improved Technology and U.S. Industry

How many times have we heard, "Well, if we can put a man on the Moon, why can't we . . . ?" Apollo is used as a standard, sometimes an inappropriate one, for what can or should be done. It is also used as an object of ridicule, of tax dollars frittered away on the Moon and its rocks, just as I am sure people complained about Lewis and Clark's bringing back a prairie dog instead of gold.

Of course none of the $24 billion that went into Apollo was spent in outer space. The bucks were spent in the United States on high-technology jobs that helped make our industries competitive in the world market. Since the Apollo days we have suffered a steady decline in our exports. Electronic products, once an American specialty, are now flooding in from Pacific Rim countries. The aerospace industry, having surpassed agriculture, is presently the number one contributor to a favorable balance of trade (around $17 billion per year), but if aerospace goes the way of electronics, our trade-deficit problems will be greatly exacerbated. The majority of our aerospace exports consist of the big Boeing and McDonnell Douglas airliners, but aviation and space technology are closely intertwined. For example, a new strong and light composite material developed for one can be used in the other. In space there is a relentless quest for improved performance, safety, reliability, and efficiency. The stringent requirements of a Mars mission will accelerate the development of technology that will be applicable to the aviation industry and to American products in general. We tire of faulty products, things that break. It's bad enough when something conks out on the ground, and it's a very serious matter in an airliner at 35,000 feet, but it's an absolute disaster months away from Earth on a Mars expedition. If one segment of American industry is forced by the very nature of its job (Mars) to achieve the highest possible standard of excellence, then that intangible attitude and the tangible products resulting from it will permeate other parts of American industry. After twenty years Apollo spin-offs are still contributing to such unlikely fields as clothing and equipment for fire fighters. In another area of concern to all of us, NASA claims that medical devices deriving from the space program have had a total economic impact of $1.8 billion since 1973. Benefits such as these are possible because of work NASA has done in computers, microelectronics, electrical power, inertial systems, computational fluid dynamics, and thermal control. . . .

Science and the Human Spirit

There is also a scientific bonanza waiting on Mars: comparative planetology, the scientists call it. They would like to study the history of the Solar System and the evolution of the planets by comparing evidence found on Earth, the Moon, and Mars. The history of Earth's climate, for example, can be partially decoded by examining ice tubes bored from the Antarctic crust. It would be extraordinarily helpful to be able to compare these with ice bored from the Martian poles. Today Earth's atmosphere seems susceptible to ozone holes, particularly a large one near the South Pole. Does Mars show similar evidence? How recent is the vulcanism on Mars? What caused it to stop? What happened to all the water that carved out deep channels? Could a similar process occur on Earth? What caused these two planets, similar in so many ways, to evolve so differently? To scientists these questions are fundamental to a better understanding of our own planet.

But even more fundamental to me than science or the pocketbook are matters of the spirit. Space exploration is a victory of technology, but only in the same narrow sense that a new thought is a victory of neurons and synapses. When we explore the Moon or Mars, we really explore ourselves and learn more accurately how we fit in. For centuries we were guided by the ideas of Ptolemy, the Egyptian astronomer who took elaborate pains to rationalize the motion of Sun, planets, and stars in terms of an Earth-centered coordinate system. Ptolemy taught that the Earth was the center of everything. Galileo set us straight and today every schoolchild has had explained that the Sun is the local center about which we and the other planets revolve. But psychologically we still cling to the Ptolemaic view, and no wonder—as long as we stay here on Earth, we are the de facto center of everything that matters. . . .

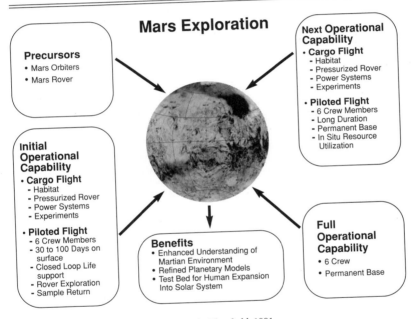

Mars Exploration

Precursors
- Mars Orbiters
- Mars Rover

Next Operational Capability
- **Cargo Flight**
  - Habitat
  - Pressurized Rover
  - Power Systems
  - Experiments
- **Piloted Flight**
  - 6 Crew Members
  - Long Duration
  - Permanent Base
  - In Situ Resource Utilization

Initial Operational Capability
- **Cargo Flight**
  - Habitat
  - Pressurized Rover
  - Power Systems
  - Experiments
- **Piloted Flight**
  - 6 Crew Members
  - 30 to 100 Days on surface
  - Closed Loop Life support
  - Rover Exploration
  - Sample Return

Benefits
- Enhanced Understanding of Martian Environment
- Refined Planetary Models
- Test Bed for Human Expansion Into Solar System

Full Operational Capability
- 6 Crew
- Permanent Base

Source: The Synthesis Group, *America at the Threshold,* 1991.

When it comes to actually colonizing the Red Planet, our Martian settlers may become fiercely proud of their locale, just as I have become of the Outer Banks of North Carolina. For the first time, the psychologists who predicted a "breakaway" phenomenon among space travelers may find themselves right. Our new Martians may not look back. They may establish ways that are different from Earth ways, they may do things better. Certainly they will have some powerful advantages. For example, the first to arrive should be a close-knit, cohesive group with no "us versus them" thinking—unless it is "us Martians" versus "them Earthlings." And with no indigenous diseases, no traditional enemies, no borders, no lack of real estate, no accumulated wealth, and a common necessity to struggle against the harsh environment, Martian settlers will not have a lot of the excess baggage that impedes friendly relations among Earthlings. They will have a fresh start, these people *of* the Earth but not *on* the Earth. All this sounds a bit utopian, I realize. On Earth, expedition members have turned violently against each other when things went wrong. Murder, rape, cannibalism—all have been recorded in small groups, along with heroic deeds. Anthropologists speak of the alpha male and the constant struggle within a group to displace him. The group purges itself, but it can be a painful and even deadly process. Humans on Mars will still have their dark side, and things won't be easy for the early settlers, but I can't help but feel optimistic about their ability to establish a colony that accepts the finest components of human behavior and rejects the basest. Selection and training of the first groups will be important in making this happen. . . .

Earth Can Learn from Mars

At any rate, our new Martians will have a variety of surroundings not available on the Moon, and some visual treats and physical sensations that may match the best Earth has to offer. Their high-tech colony will have to grow using un-Earthly techniques. Their rivalry with Earth, their admiration or criticism of Earth—any of these consequences of a human presence on Mars can only help the home planet, help it to understand itself better and to write a prescription for change. If it is true that someone who doctors himself has a fool for a patient, then perhaps a planetary consult might be in order, a second opinon from a very different place.

I don't mean for Mars to be an escape valve. For more than a hundred years we have debated the theory of Malthusian expansion, and more recently the Club of Rome has attempted to define some limits to growth. On Earth the supply of food may always lag behind the production of new human beings, but Mars cannot be counted on to alter this equation, either by supplying resources to Earth or by siphoning off Earth's excess population. The distance is too vast, the cost too high. But living in close quarters under a dome, squandering nothing, recycling products to the maximum, Martian colonists may very well find new antidotes to Malthusian pressure.

I had to see a second planet—the Moon—to fall in love with the first. Perhaps we need to live on a third—Mars—to go beyond love to a successful marriage with the first. How would people on Earth react without any weapons, for example? Since the dawn of history our hunting ancestors have had them, and today, as Carl Sagan has written, "The U.S. and the U.S.S.R. have now booby-trapped the planet with nearly 60,000 nuclear weapons." There would be no need to carry weapons to lifeless Mars, and it would be easy to prevent their transport from Earth. Therefore, for a while anyway, a modern society would be totally without them. What could we Earthlings learn from this? How would Martian disputes be settled? In my opinion, escaping from hydrogen bombs is not a valid reason for leaving Earth, but learning to live without them may be.

Mars: The Next Step

Planets are all we have left to explore, in a physical sense, except for a bit of roaming on the sea bottom. Only the inner planets are accessible now. Mercury and Venus are impossibly hot, and we have already done the Moon (wow, will some of my compatriots argue with that!). Therefore, if we have a spiritual need for a new frontier—and I believe we do—Mars is it. If there is a migratory drive within us—and I believe there is—it will lead us to Mars. If there is an extraterrestrial imperative, Mars is surely the next logical stepping-stone on the endless journey to the stars. Our bodies are no more than star stuff that coalesced along with the Earth, debris from the original explosion that created everything we know. We won't stay here. Call it genes, character, culture, spirit, ethos: by whatever name, it is within us to look up into the night sky and be curious, within us to commit our bodies to following our eyes.

Exploration: I don't want to live without it; I don't want to live with a lid over my head. What will exploration gain us, beyond allowing us to dream? T.S. Eliot said it better than I can: "We shall not cease from exploration and the end of all our exploring will be to arrive where we started and know the place for the first time."

"Sending people to Mars will be romantic but ridiculously impractical."

The U.S. Should Not Explore Mars

Gregg Easterbrook

Cost estimates of a space flight to Mars range from tens of billions of dollars for an unmanned mission to hundreds of billions of dollars to land a crew. In the following viewpoint, Gregg Easterbrook argues that these costs make a mission to Mars prohibitive. He charges that both a bureaucratic NASA and aerospace companies preoccupied with program contracts will be unable to reduce the costs of a Mars mission. Easterbrook is a contributing editor for *Newsweek*, the *Atlantic*, and the *Washington Monthly.*

As you read, consider the following questions:

1. Why will it be so expensive to send astronauts to Mars, according to Easterbrook?
2. In the author's opinion, why does NASA favor the space shuttle for many of its launches?
3. According to Easterbrook, why does NASA care little about meeting deadlines and lowering costs?

George Bush wants the United States to go to Mars: he's given several speeches on the subject. Dan Quayle, whose position as director of the National Space Council is one of his few real responsibilities, is very big on the Mars thing. Even Richard Darman, in a speech during which he ridiculed everybody else in Washington for refusing to get tough on federal spending, nevertheless endorsed a blank check for a Mars expedition, declaring it would be "romantic." Depending on who you're locked in the capsule with, perhaps.

Foolish Mission

Why this sudden red planet chic? Travel to Mars makes an incredibly zoomy topic for nonbinding speeches, so it's attractive to administration officials looking to sound high-tech, or desiring an oratorical diversion from intractable terranean problems. Any suggestion of a major new space push delights the big aerospace contractors, who constitute a prime Republican constituency. A Mars flight also engages the enthusiasm of a small but significant segment of the voting public. If only it weren't such a silly notion.

Expeditions to Mars are going to occur someday; I am reasonably optimistic that one will occur during my lifetime. But then, according to mortality tables, I should live till the year 2029. In the decade or two to come, sending people to Mars will be romantic but ridiculously impractical.

In his book John Noble Wilford, an accomplished science writer for *The New York Times*, does an excellent job of spelling out the alluring aspects of a potential red-planet mission: the scientific enticements of the objective, the likely drama of the transit, Mars's place as a beacon for human aspirations. As a sales pitch, it's a fine book. Intellectually, however, *Mars Beckons* is shallow and slothful. Wilford basically skips one of the central questions of Mars exploration proposals—why people should go when robot spacecraft could accomplish all immediate scientific objectives for a fraction of the expense or risk. And on the key reality-check issue, money, *Mars Beckons* is silent. This 194-page volume has but a single paragraph on the question of cost.

Consider that the basic price tag for the space station grew from $8 billion to $32 billion since 1985, even though the size of the facility shrunk. What does this suggest about what the true cost of Mars exploration might be?

Cost of a Mars Flight

Thirty-two billion dollars for the space station will buy a few modules strung together just 250 miles from their launch pad: with no main engines, no landing or ascent craft, and no crossing of new technological frontiers such as closed-cycle life support, since supply ships will arrive regularly. A Mars mission would require ships equal to dozens of space-station modules: self-propelled and self-sustaining, able to travel not 250 miles but 48 million miles and back over a period of two to three years.

Steve Benson. Reprinted by permission: Tribune Media Services.

Mars-bound ships will require much more radiation shielding than do spacecraft in Earth orbit, and probably simulated gravity, as so far there are no known palliatives for the muscle loss, bone resorption and extreme productivity drop-off Soviet cosmonauts have experienced in stays of about a year aboard the Russian *Mir* orbital laboratory. The expedition would require such complexities as a fully equipped surgical theater (if, say, an eight-person crew is gone for three years, statistically, someone's going to require an operation). A highly advanced degree of safety redundancy would be advisable—probably, as with Columbus's three-ship convoy, a complete extra vessel—because, unlike the space station, the Mars convoy would be far beyond the reach of emergency assistance. Landers much larger and more powerful than those of the Apollo mission would be required for the ride to the Martian surface. In short, a Mars fleet would make the $32 billion space station—which no one in the Bush administration knows how to fund—seem like a Mattel toy.

(Note to fans of dubious space station justifications: Not only does the design now include an unmanned "free flying" platform next to the station, where scientific experiments can be isolated from the disruptive effects of astronauts shaking the structure with their movements, but NASA is furiously trying to engineer robots that would handle much of the "spacewalk" assembly and repair work. So just what is the space station crew going to do? Monitor telemetry from the platform and the robots, which could, with current technology, be done as easily from Dan Quayle's living room.)

Launch Costs

Fuel-launching costs alone guarantee that a Mars mission will be exceedingly expensive. Since it's not going anywhere, the space station will require only incidental supplies of rocket fuel. A 1988 report by the former astronaut Sally Ride estimates that an austere mission to Mars—employing one cargo ship and one crew ship (no survival redundancy)—would require 2.2 million pounds of propellants at departure from low-Earth orbit. Ride didn't assess what that means, and neither does Wilford. The current space shuttle true payload cost to low-Earth orbit is in the neighborhood of $7,000 per pound. At that price it would cost about $15 billion—more than NASA's current annual budget—just to launch the fuel for a single, stripped-down Mars mission.

From a technical standpoint, the principal barriers to Mars travel are the same today as they were when the Apollo moon program ended. Two major technical breakthroughs are required.

One breakthrough would be a dramatic reduction in the cost of putting pounds into low-Earth orbit, where a Mars-bound craft would be assembled. As long as NASA clings to its insistence that the excessively expensive and dangerous shuttle is the answer to all space-launching questions, grand ambitions like Mars landings or Moon colonies will remain fiscal pipe dreams. As is, shuttle prices are so high that NASA is having trouble getting planetary probes and communication satellites in orbit, objects with a tiny fraction of the weight necessary for serious manned space flight.

Shuttle Alternatives

There are alternatives to high shuttle prices, but NASA for institutional reasons resists them. The current-technology alternative would be a system of relatively low technology, expendable, "big dumb boosters" for launching cargo, combined with a "spaceplane" similar to a small shuttle, air-launched from a carrier aircraft like a 747, for moving people to orbit.

Many scientists have suggested the U.S. switch from the shuttle to such vehicles, and NASA has always growled. With hardware like this the vast majority of launches would be unmanned, whereas with the shuttle a crew goes along for every routine relay satellite launch, engaging huge cost and risk but ensuring a continued flow of appropriations to NASA's existing hierarchy of contractors and manned-flight centers.

A Foolish Dream

Men and women surely will journey to Mars someday. For now, anyone who advocates Mars travel is living in a dream world. NASA has put the cost of a Mars landing at $400 billion, meaning the actual figure might be twice as high. A NASA study estimated that an "austere" Mars mission would require 2.2 million pounds of propellant at departure from Earth orbit. At current shuttle prices, merely launching that quantity of fuel—to say nothing of financing the Mars ships—would cost $25.7 billion, almost twice NASA's annual budget.

Gregg Easterbrook, *The New Republic*, July 8, 1991.

The second breakthrough requirement for a practical Mars mission is a propulsion source other than standard chemical rockets, so that either the fuel volume required at departure from orbit is dramatically cut back or the Mars transit time is reduced, alleviating the need to carry huge caches of supplies and diminishing the psychological and physical strains that will attend spending years in zero gravity and very close confinement.

Nuclear Engines

The leading current-technology candidate for a new propulsion source is the nuclear rocket engine, which was researched but never flight-tested for the Apollo program. Nuclear engines require far less fuel mass than chemical rockets, and can fire for much longer periods, which means more speed. Operating beyond low-Earth orbit where the natural background radiation from solar and cosmic rays is already at a fatal level, nuclear-powered Mars craft would pose no threat to the environment. Nuclear engines might be the answer for practical interplanetary craft, but not even National Space Council buffs seem willing to try to explain this to the public, for fear of the "n" word. Since the reactor core of a nuclear rocket would produce current rather than thrust, references are usually disguised with the euphemism "electric propulsion."

(Another note: though nuclear-propelled spacecraft could attain higher speeds than possible with conventional chemical rocket engines, they would take days or weeks to accelerate, making such propulsion of limited interest to military engineers. Knowing that nuclear rocket power would have mainly civilian applications, the Air Force is already doing general research into propulsion from antimatter, especially antiprotons. . . .

Preliminary speculation is that mid-21st century antimatter technology may enable the construction of machines moving at about 1 percent the speed of light—nowhere near fast enough for exploration of the closest star but plenty fast for military operations among this solar system's planets or dogfights in orbit.)

NASA's Inefficiency

Oh yes, NASA also protests that fundamental space program changes, like a switch to big dumb boosters for cargo and small spaceplanes for people, would require a multibillion dollar R&D investment. This is unquestionably true. But once that investment was made, such systems would enable annual space operating costs to decline. Somehow NASA considers development of cost-cutting launch systems a wildly unreasonable use of taxpayer money, yet it is perfectly happy to talk about a Mars landing—whose full price some NASA internal documents now put at $500 billion.

Prices like that (about three years' worth of the current federal deficit) reflect the daunting nature of any serious Mars expedition and show cultural change in the way NASA and its suppliers do business. During the Moon push, NASA was a progressive, can-do agency, more interested in results than in "process" and bureaucratic preservation. NASA contractors reflected this progressive attitude: unbilled overtime rather than cost-plus foot-dragging was the rule. NASA and its suppliers during the Moon days stood for the kind of government program people could believe in.

Today, the mind-set of the defense-industrial complex has taken over. NASA managers care less about how many months or millions anything takes than about preserving budgets and staff levels. Aerospace contractors in turn mirror their client's attitude, devoting more creative energy to lobbying and turf protection than technical innovation. The contractors feel that as long as the space program remains locked in wheelspin they might as well run up the bill, and from their standpoint there is a certain sad logic to this analysis. Since 1988 NASA's budget has risen 36.6 percent, but practically all the money has gone to cost increases in existing programs, not new projects.

Cost, practicality, and NASA bureaucratic coagulation—not the niftiness of space travel, which everybody's already convinced of—are the subjects on which light must be shed before Mars flight can become any more than a rhetorical goal.

"Space Station Freedom *should be viewed as a research and development investment in our future."*

The U.S. Should Build Space Station *Freedom*

NASA

Proposed by Ronald Reagan in 1983, Space Station *Freedom* is scheduled to be completed in 1999. In the following viewpoint, NASA argues that *Freedom* can generate scientific innovations, new products, and jobs that will benefit science, industry, and the economy. NASA maintains that such innovations in technology and materials will enhance U.S. industries' competitiveness. Also, NASA believes that *Freedom* will demonstrate America's technological expertise to the world and promote international sales of U.S. aerospace products and services. NASA, headquartered in Washington, D.C., is the government agency responsible for space exploration activities. The viewpoint is the written testimony NASA submitted to Robert Kerrey, a Democratic senator from Nebraska.

As you read, consider the following questions:

1. How can research aboard *Freedom* benefit long-term, manned space exploration, according to NASA?
2. According to NASA, why are aerospace industries so important to the U.S. economy?
3. In the author's opinion, how will *Freedom* strengthen America's leadership in space?

From NASA's testimony submitted to Sen. Robert Kerrey (D-Neb.) for a Senate Appropriations Committee hearing, May 8, 1991.

K*errey*: A number of scientists have criticized the downsized space station, questioning its scientific value. What do you see as the major scientific benefit of the station?

NASA: Space Station Freedom will provide the ability for world-class research to be conducted in the areas of Space Life Sciences and Microgravity Sciences.

Space Station Freedom (SSF) has great significance for NASA Space Life Sciences Programs. For the first time it will be possible to replicate experimental data with appropriate controls and real-time analytical capabilities over extended periods of time. SSF will provide the means to acquire basic knowledge on mechanisms of gravity perception while paving the way for extended-duration exploration missions with humans.

Spaceflight Research

Research in long-duration human life sciences is considered to be an essential objective of SSF's pressurized modules. Space Station fulfills the Space Life Science goals to develop the knowledge required to assure crew health, well-being and performance by studying adaptation of living organisms to spaceflight. It also will conduct fundamental scientific investigations on the role gravity plays on the life process itself.

A large majority of NASA's Space Life Science research requires the extended operations of a fully-equipped, permanently staffed laboratory. Such a facility enables the life sciences research community to perform biomedical, biological life support and exobiology studies. First, this allows exposure of experimental specimens over a significant period of the organisms' life span, in some cases for multiple generations. Secondly, it permits systematic research on statistically significant specimen populations with real-time experiment replication to validate initial experiment observations. Finally, controlled studies utilizing matched specimen populations on the centrifuge will determine threshold responses and dose-response relationships in a wide range of living systems before the requirement for a human-rated centrifuge is validated.

The overarching objective of the NASA Microgravity Program is the use of space as a laboratory to conduct basic research and development. The on-orbit microgravity environment with its substantially reduced buoyancy forces, hydrostatic pressures, and sedimentation, enable us to conduct scientific investigations not possible on Earth. This environment allows processes to be isolated and controlled and measurements to be made with an accuracy that cannot be obtained in the terrestrial environment. Many of the processes being studied also play dominant roles in diverse Earth-based technologies as well as technologies that support NASA's overall goals for future extraterrestrial exploration. The low-gravity environment therefore demands investigation both for scientific and technological reasons.

NASA has defined three major science categories in order to develop a program structure. These areas of research and development to be conducted in Space Station encompass 1) fundamental science, which includes the study of the behavior of fluids, transport phenomena, condensed matter physics and combustion science; 2) materials science, which includes electronic and photonic materials, metals, alloys, glasses and ceramics; and 3) biotechnology, which focuses on macromolecular crystal growth and cell and molecular science. Experiments in these areas typically seek to provide observations of complex phenomena and measurements of physical attributes with precision enhanced by the microgravity environment. Microgravity research can also provide insights that could lead to a better understanding of Earth-based industrial processes and/or space-based production of new materials with unique properties.

"Did they have to scale back the space station THIS MUCH?!"

Bruce Beattie. Reprinted with permission.

The goal of microgravity research on the SSF during the early years will be to maximize, to the extent practical, the quality of the science return. As the research matures, the experiments will become more sophisticated and the program will define and cultivate focused research areas of the highest scientific potential. Achievement of these goals will be accomplished through implementation of a plan that will concentrate on research in the areas of fluid dynamics and transport phenomena, combustion science, and biotechnology during the utilization flights and an emphasis on materials science and protein crystal growth during the periods between the Shuttle flights.

There are six facilities planned for utilization of the SSF on-board environment for microgravity experiments. They are as follows: Advanced Protein Crystal Growth Facility, Space Station Furnace Facility, Modular Containerless Processing Facility, Modular Combustion Facility, Fluid Physics/Dynamics Facility, and Biotechnology Facility.

Contribution to Science and Industry

Kerrey: In what specific areas do you believe it contributes to U.S. competitiveness?

NASA: The Space Station Freedom program drives the development of many technologies. Examples of technologies and systems that will be advanced by Space Station development and operation include: environmental-control and life support; power generation, storage and management; thermal control; crew health care; data processing and distribution; guidance, navigation and control; structures and materials; and operational methods. Commercial applications for these technologies could have strong market potential which will enhance U.S. industrial competitiveness.

The Space Station program contributes to U.S. competitiveness in other ways. The Space Station program has created thousands of engineering and technical jobs and has replenished the pool of knowledge and experience in the design, manufacture and operation of numerous technologies. Space Station Freedom has provided inspiration and will continue to provide inspiration to our children to make science and technology their life's work. The Space Station program will demonstrate the United States' ability to lead the world in cooperative technology development, contributing to our prestige and influence in these matters.

Construction of Space Station Freedom will provide U.S. industry with otherwise unattainable expertise in the construction and assembly of large facilities in space, a key prerequisite to private sector, large-scale investment in space and a significant enhancement to U.S. industry's competitive position around the globe. . . .

Kerrey: What would funding the station do for the advancement of science and U.S. competitiveness that additional funding for space science or additional funding of National Science Foundation individual researchers would not do?

NASA: The U.S. supports a broad range of science activity, both ground- and space-based, to ensure the advancement of science and U.S. competitiveness. Just as the National Science Foundation provides the infrastructure to support its ground-based science, so must NASA through laboratories such as Space Station Freedom for space-based science. Consequently, Space Station Freedom is viewed as a critical element in a well-balanced U.S. science program.

A Space Milestone

Space Station Freedom is a major milestone in our planning for the future. This permanently occupied orbiting base will help to maintain U.S. space leadership into the 21st century. It will play a vital role in science, exploration, and space commercialization. . . .

Freedom offers a unique facility for developing new technologies, products, and processes. Among the exciting possibilities are the development of new pharmaceutical methods and processes and treatments for serious diseases.

Research on Freedom will help prepare humans for the long-duration space missions of the Space Exploration Initiative.

National Space Council, *1990 Report to the President.*

Space Station Freedom is a visible demonstration to the world of our technological prowess and expertise which is of benefit in stimulating international sales of U.S. aerospace products and services. The technologies under development for the Space Station systems significantly extend, or in some cases, establish the leadership of the U.S. industry. Space Station Freedom should be viewed as a research and development investment in our future and as a provider of goods and services to other nations that choose to utilize space-based systems. Our current positive balance of aerospace trade is predominantly associated with commercial aircraft sales. This position of leadership was forged with government leadership that focused the entrepreneurial spirit of the private sector. The market for space systems and services is growing and will continue to do so for decades. The Space Station program will provide U.S. industry with technology and experience that will ensure U.S. competitiveness in the next century.

Additionally, once in orbit, Space Station Freedom will provide unprecedented opportunities for first-class basic and applied research in life sciences and microgravity materials research. The research will be directed toward both our commitment to improve the quality of life on Earth and our national goal of world leadership in space development.

Kerrey: If the station were not funded, would you view that as essentially the end of manned space exploration?

NASA: Cancellation of the Space Station Freedom program would mean a total disruption of the United States' manned space program. It would have a devastating impact upon the unique and highly refined institutional base resident at NASA and its contractor/university teams. The current NASA program is very interdependent. Space Station Freedom does not stand alone. It is woven into the fabric of our space science, space transportation and space tracking programs. Cancellation of Space Station Freedom would place in danger the future of the New Launch System, the Advanced Solid Rocket Motor, and our microgravity and life sciences research programs, to name just a few.

Manned space flight will nonetheless continue around the world. The Soviets have a strong track record on manned space flight activities, with no reduction in sight, and both the Europeans and the Japanese have expressed a long-term interest in developing autonomous manned space programs. The question is not whether manned space exploration will continue, but rather, who will be the world leader, and whether the U.S. will be involved.

Space Station Alternatives

Kerrey: If we returned to a Skylab or an Apollo-Soyuz type project, could we continue to gather significant information on manned space activity or have we simply done what we could at that level?

NASA: A Skylab type space station would fall far short of Space Station Freedom in terms of both quality and quantity of research capability. For example, our previously flown Skylab had only half the interior volume and one-seventh of the power as compared to the permanently manned Space Station. Additionally, the cost to develop and deploy a Skylab-type design Space Station would be prohibitively expensive when the cost of building the module(s) as well as the cost of developing and producing a launch vehicle to place it in orbit are considered. A Skylab-type design could provide information on the effects of extended duration weightlessness on humans.

The Apollo-Soyuz project was a short-duration project which would provide no benefits that are not currently obtainable from the Space Shuttle with Spacelab.

Kerrey: If the space station were terminated, what would you see as the priorities and principal focus of NASA in the years ahead?

NASA: We have no plans for the future that do not include Space Station Freedom. As stated in a previous answer, the Space Station is woven into the fabric of nearly all of our current and planned programs.

"*Space Station* Freedom *is neither well suited for scientific research nor for space exploration.*"

The U.S. Should Not Build Space Station *Freedom*

Bruce C. Murray

The proposed Space Station *Freedom* has been criticized by scientists and others for both its purpose and its estimated $30 billion cost. In the following viewpoint, Bruce C. Murray argues that the space station will hamper exploration because it is unsuited for its scientific tasks. Murray contends that a human presence on the station will interfere with tasks that could be better accomplished by an unmanned mission. He also believes that diverting limited resources to the building of the space station will sacrifice other worthwhile space programs. Murray is a professor of planetary science at the California Institute of Technology and vice president of The Planetary Society, an organization that promotes space exploration. Both organizations are in Pasadena, California.

As you read, consider the following questions:

1. According to Murray, how do NASA's problems threaten the scheduled completion of *Freedom*?
2. In the author's opinion, why will construction of *Freedom* threaten astronauts' lives?
3. According to Murray, how could a more productive space station than *Freedom* be built?

From Bruce C. Murray's statement to the Senate Subcommittee on Science, Technology, and Space, April 16, 1991.

The Planetary Society represents over 100,000 members who strongly support planetary exploration. In the mid-1980s, we pioneered public advocacy for human flights to Mars as the centerpiece of what is now called "Mission From Planet Earth." And we have consistently argued that "A Space Station Worth the Cost" is a necessary step in that endeavor. But Space Station *Freedom* (SSF) will not be a stepping-stone to the planets: In reality it is proving to be an impediment to both human and robotic exploration. In its current incarnation, Space Station *Freedom* is neither well suited for scientific research nor for space exploration. Its projected costs and schedule are not credible. It is a recipe for a programmatic disaster that will last well into the next century.

The National Research Council's Space Studies Board (SSB) has clearly and forthrightly spelled out why Space Station *Freedom* will not fulfill priority science goals. The SSB represents all the space sciences—biology, astronomy, planetary, solar, microgravity, materials processing, life sciences. In its March 1991 report, the Board concluded that Space Station *Freedom* "does not meet the basic research requirements for the two principal scientific disciplines for which it was intended"—microgravity materials processing and life sciences.

Inadequate Research

In particular, the human presence will disturb microgravity materials processing. Even routine station operations, which include the mating of modules and docking of shuttles, will hamper long-duration experiments. Microgravity experiments have always been best carried out either in totally automated fashion or on human-tended platforms specifically designed for that purpose. That remains the correct—and affordable—direction for the U.S.

Even more serious for the future of U.S. human space exploration, in the present SSF program there will be no significant life-support research in space until the next century. Americans will make no long-duration flights until the next century—well over a decade after the Soviets reached the one-year-in-orbit milestone. There will be no variable gravity studies and only primitive experiments with artificial gravity. There will be no work on closed ecological systems with humans in the loop. There is no capability to develop and test interplanetary transportation and propulsion systems. But all these omissions are essential preparation for one day sending humans to the planets.

It is more apparent than ever that microgravity and biomedicine are mutually exclusive, and should be accomplished with separate, dedicated platforms, perhaps on different schedules.

101

The White House Office of Science and Technology Policy (OSTP) itself concluded that neither the space station's cost nor the effort to build it was justified scientifically. They recommended that NASA refocus the project solely on research to "support human life during long space flights." This research is necessary if we are to accomplish the goal offered by President Bush in his July 20, 1989 speech setting forth a Moon-Mars initiative. The Planetary Society deplores NASA's sluggish response to the challenge of space exploration put forward by the Bush Administration.

THE BLIND MEN AND THE WHITE ELEPHANT

© 1991, John Trever, *The Albuquerque Journal*. Reprinted with permission.

It is not only the scientific community and those who pursue human exploration of the solar system who reject this space station as neither useful nor cost effective. No private or military users have plans for it either.

Indeed, why build this 1980s-oriented station for which there is no 1990s user community? The question, "Who will use the space station?" has plagued NASA ever since President Reagan first approved it in late 1983. The search for users continues, but no credible community has emerged.

In 1990 the Advisory Committee on the Future of the U.S. Space Program—the Augustine Committee—gave as their first recommendation that "the civil space science program should have first priority for NASA resources." The National Space Council, the scientific community, The Planetary Society, professional publications, major newspapers and the Vice President strongly endorsed these findings.

One might therefore have expected SSF to be redesigned to fulfill a primarily scientific mission. In 1990 the U.S. Congress took the extraordinary step of intervening in the program by directing NASA to reconfigure the space station into a more modular, more robust, and more affordable form. NASA maintains that their latest redesign is responsive to Congressional interests and the Augustine Committee's recommendations. It is not. Widespread skepticism and concern continue, as expressed by the SSB and OSTP statements.

Then, on March 19, 1991 Vice President Quayle acknowledged in a letter to the NASA Administrator that science wasn't the primary purpose after all. The space station should proceed because, "the ultimate mission [of the space station] is necessary to the reaffirmation of the leadership in space of the United States of America, the world's only superpower."

Unfortunately, the Vice President's letter didn't explain how a station that won't be ready until 1999 at the earliest, with no U.S. habitat module nor artificial gravity capability, will reaffirm American leadership in space. Nor did it make clear how SSF will dramatically extend the achievements of the highly successful American *Skylab*, which were accomplished 25 years earlier. Nor did it explain how SSF in 1999 will leapfrog accumulated Soviet accomplishments with *Mir*, which has been in orbit since 1986.

Problems with NASA

Sadly, it is difficult to escape the conclusion that the most compelling motivation for much of the current support for SSF is anxiety over the institution of NASA and the perceived consequences to the agency of further delay. The President's July 20, 1989 call for a resumption and extension of human space exploration seems to have been overtaken by a Potomac fog.

However much compassion one may feel for a struggling agency, Congress must fully comprehend and accept the consequences of proceeding now with an expensive, open-ended and backward-looking human space program. Several years after SSF was proposed, and with one-third of the estimated total funds already appropriated, SSF remains years from full operation. Further delay is almost inevitable. Total costs will increase beyond current estimates. The current scaled-down, redesigned station still requires at least 23 successful, on-schedule shuttle flights and about 500 hours of EVA [extravehicular activity] construction. Both of these requirements are well beyond demon-strated U.S. capability. Every other NASA program (except the shuttle, which is required for SSF) will be squeezed mercilessly. The vision enunciated by President Bush on July 20, 1989 will be the first casualty. But future breakthroughs in space astron-omy, in planetary exploration and in monitoring Earth from space will become hostages as well.

Poor Design and Purpose

It's sometimes suggested that those of us who criticize Freedom's present design are just ticked off because the station is not going to be equipped with all the bells and whistles we want. That's wrong. Believe it or not, we want to spend *less* money, but spend it more wisely. We need to forget the rush to get "something" up there as quickly as possible and do the job right—either a cheaper unmanned platform for microgravity experiments alone, or a more sophisticated station for research on human space ex-ploration. Freedom is too big for one of its jobs, and too small for the other.

Robert F. Sekerka, *The Wall Street Journal*, June 10, 1991.

Giving SSF the go-ahead now will demand the sacrifice of many other highly meritorious NASA programs. . . . We will have slowed our preparation for eventual human exploration. And we will have delayed building a new, U.S. heavy-lift launch vehicle (HLLV), essential for human exploration beyond Earth orbit. SSF is blocking, not advancing new human exploration of the solar system.

If, on the other hand, an HLLV were built *before* the space sta-tion, then it could play the role for SSF that *Saturn 5* did for *Sky-lab*. We could then launch a large, prefabricated and truly sig-nificant biomedical facility, probably in about the same time frame as the current SSF program will reach permanently-manned status. This approach was elucidated in a Planetary So-ciety workshop in Washington on September 19, 1990. Now, however, the funds for HLLV development will be in direct competition with those for the expensive, shuttle-based Space Station *Freedom*.

Risks of Spaceflight

Finally, we must never forget *Challenger's* traumatic lesson—that human spaceflight has costs beyond dollars. *Challenger's* sad legacy was to remind us that humans should *not* be sent into space except for purposes worth risking lives for. SSF involves 23 to 26 cargo-carrying shuttle trips, some 500 hours of EVA, and novel human participation in on-orbit assembly, all of which constitute substantial hazards. The Congress must be convinced that the purposes of SSF are worth some loss of life, because that is certainly a plausible outcome. In comparison, three *Apollo* astronauts died in preparation for the Moon journey, and three others nearly perished on the journey itself. The American public accepted these losses and risks as justified by the historic achievement of the first human flight to the Moon.

In summary, The Planetary Society advocates a space station that aims well beyond *Skylab* and *Mir*, that will open the way to human exploration of the Moon, Mars and nearby asteroids. We strongly advocate advancing U.S. capability for exploring the planets incrementally in a balanced program. We encourage a global, cooperative program for both Mission From and Mission To Planet Earth. It is distressing, after the bright promise of President Bush's Space Exploration Initiative and the wise Augustine Committee Report, to see America slipping back into the quagmire of an inadequate program, oriented towards symbols rather than historic achievements, motivated primarily by parochial institutional needs, and lacking a commitment to genuine exploratory accomplishments.

Important Choice

The choice is whether or not to proceed with a space station that is not worth the cost, puts humans at risk for mundane purposes, does not meet the goals of those who would use it, portends a financial and programmatic disaster for the civil space program, and is being funded primarily because of narrow bureaucratic concerns. The U.S. space program could instead truly commit itself to an era of new historic achievement, one that will ultimately put humans on another planet, will map the heavens in unimagined detail, and will affordably and promptly help all of us manage Planet Earth. Space Station *Freedom* will delay, not accelerate, these exciting objectives.

Ranking Priorities for the U.S. Space Program

In this chapter you have read several different opinions on which programs the U.S. space program should pursue. This activity will allow you to explore the priorities you think are important for U.S. space exploration. Your answers may differ from those of other readers, mirroring the controversy concerning the value of space exploration.

Opinions about what the space program's priorities should be may vary among people. For example, while the primary objective of an advocacy group such as the National Space Society is to promote manned space exploration, a university researcher may seek unmanned exploration for its benefits to science. Similarly, a member of a task force on poverty may object to the expense of space programs, arguing that problems such as unemployment and malnutrition deserve higher priority.

Part I

Working individually, rank the concerns listed below. Decide what you believe to be the most important priorities for U.S. space programs and be ready to defend your answers. Use number 1 to designate the most important concern, number 2 for the second most important concern, and so on.

_____ using satellites to collect solar energy

_____ landing astronauts on Mars

_____ researching space technology only if it can benefit life on earth

_____ building a space station

_____ constructing cities on Mars

_____ building more efficient space vehicles

_____ studying earth's environment from space

_____ developing space tourism

_____ sending unmanned probes into space

_____ deploying space weapons

Part II

Step 1. The class should break into groups of four to six students. Students should compare their rankings with others in the group, giving reasons for their choices. Then the group should make a new list that reflects the concerns of the entire group.

Step 2. In a discussion with the entire class, compare your answers. Then discuss the following questions:

1. Did your opinion change after comparing your answers with those of other group members? Why or why not?
2. Consider and explain how your opinions might change if you were:
   a. chief executive officer of a NASA subcontractor
   b. head of an environmental activist group

Periodical Bibliography

The following articles have been selected to supplement the diverse views presented in this chapter.

Aerospace America	"Face to Face with Bill Green," January 1992. Available from the American Institute of Aeronautics and Astronautics, 370 L'Enfant Promenade SW, Washington, DC 20024.
Oleg Borisov	"Why Mars Should Be Earth's Next Goal," *Air & Space,* February/March 1990.
William J. Broad	"The Shuttle Saga May Be Heading Toward a Close," *The New York Times,* December 16, 1990.
Shawn Carlson	"Virtual Mars?" *The Humanist,* March/April 1991.
Leonard David	"NASA's Countdown to Nowhere," *Ad Astra,* October 1991.
Final Frontier	"Cosmic Ray: A Talk with Ray Bradbury," January/February 1991. Available from PO Box 534, Mt. Morris, IL 61054-7852.
Brenda Forman	"Voyage to a Far Planet," *Omni,* July 1990.
Bill Green and Bill Nelson	"A House Divided," *Omni,* July 1990.
Bill Green and Robert Roe	"Should Congress Support Manned Space Exploration?" *American Legion,* July 1990. Available from 5561 W. 74th St., Indianapolis, IN 46268.
Jerry Grey	"It's a Bleak Future Without Freedom," *Aerospace America,* June 1991.
T.A. Heppenheimer	"Beyond Tomorrow," *Reason,* May 1991.
Christopher P. McKay and Robert H. Haynes	"Essay: Should We Implant Life on Mars?" *Scientific American,* November 1990.
Eliot Marshall	"The Shuttle: Whistling Past the Graveyard?" *Science,* October 1990.
Robert G. Oler	"More Shuttles, Please," *Ad Astra,* December 1991.
Donald F. Robertson	"Mars on the Cheap," *Final Frontier,* November/December 1991.
Carl Sagan	"Why Send Humans to Mars?" *Issues in Science and Technology,* Spring 1991.
Harrison Schmitt	"Justifying Freedom," *Final Frontier,* March/April 1992.
Robert F. Sekerka	"The Space Station—A $30 Billion Boondoggle," *The Wall Street Journal,* June 10, 1991.

Should NASA Be Eliminated?

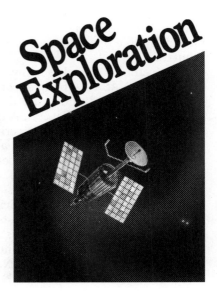

Chapter Preface

During the early years of America's space program, NASA enjoyed much success. In 1962, less than four years after the agency's inception, astronaut John Glenn became the first American to orbit the earth. Seven years later, Apollo astronaut Neil Armstrong became the first human to walk on the Moon. With such historic feats, Glenn and Armstrong instantly became heroes, and NASA became a symbol for American ingenuity.

But in recent years, NASA has found similar milestones difficult to accomplish and has been beset with problems. The greatest tragedy came in 1986, when the space shuttle *Challenger* exploded, killing all seven astronauts aboard. This was followed by the disclosure of gross negligence by NASA and its subcontractors. The *Challenger* fiasco and the 1990 launching of the flawed Hubble Space Telescope have severely tarnished NASA's reputation.

These setbacks have led many observers of the space program to question NASA's effectiveness and even to debate whether NASA should continue to be the only organization exploring space. Critics charge that NASA is plagued by bureaucracy and mismanagement. In addition, critics argue that as part of the government, NASA constitutes a monopoly by being the sole space-exploring agency. More vocal opponents argue that NASA should be abandoned altogether and that space exploration be turned over to private companies.

NASA hopes to allay such criticism with a return to the Moon and a mission to Mars early in the next century. A journey to Mars could duplicate the success of the first Apollo landing and restore the prestige that NASA once held. The authors in the following chapter debate whether NASA should be allowed to pursue these goals.

"The resiliency, innovation, and skill characteristic of NASA has persevered."

NASA Can Competently Explore Space

John Lawrence

NASA's space program is renowned for its achievements, such as the first lunar landing, the *Voyager* space probes, and numerous space shuttle missions. In the following viewpoint, John Lawrence argues that NASA can still effectively explore space, despite much unfair criticism, because the agency has the expertise to complete its exploration goals. Lawrence contends that NASA will continue to achieve its objectives, such as lunar colonization and Mars exploration, just as it kept its promise to land astronauts on the moon. Lawrence is a legislative affairs specialist at NASA headquarters in Washington, D.C.

As you read, consider the following questions:

1. Why does Lawrence believe that much of the criticism of NASA is unfair?
2. In Lawrence's opinion, why does NASA stress the need for sophisticated launch systems?
3. According to the author, how do NASA programs benefit the economy?

Adapted from "The Demythification of NASA" by John Lawrence, *NASA Activities*, November/December 1990.

It's axiomatic—perception is more important than fact. And if that isn't sufficiently bizarre, consider the new law which has lately come into play in some circles. It should be known as "the NASA Corollary"—fact and opinion may be not only unrelated, but at polar extremes.

One explanation for the phenomenon is something akin to the old Indian adage that your greatness is measured by the strength of your enemies. The legion of self-described adversaries and watchdogs may reason that if they're wiser than NASA, then they are wise indeed. It's flattery of a perverse sort, that such self-aggrandizement is at NASA's expense. What is particularly perplexing, however, is that the ridicule is generally in the form of opinions that come from partial information, misinformation, or (there's no sweet way to say it) absolute ignorance of the facts.

Opinions and Attitudes

Given the aforementioned axiom, the effect of such commentary is worrisome. NASA operates in a climate which requires the confidence of the White House, Congress, the public, and international partners. The extent to which this hostile barrage affects the space program is difficult to measure. Although an admittedly imprecise science, the literature of the communications field is remarkably consistent in its assessment of how opinions and attitudes are shaped. No evidence suggests that journalists or public figures have a direct impact on the public conscience. Rather, attitudes tend to be influenced by a group loosely identified as "opinion leaders." It is an ethereal group, never precisely defined. The prevailing notion is that opinion leaders are almost always personal contacts such as friends or family members, who have first, second, or third-hand sources into an issue.

By inference, the extended circle of NASA/contractor employees, their families and friends, can be opinion leaders on issues regarding the space program and can be influential in shaping attitudes. Luckily, the misperceptions they must repudiate do not stand up well against the facts. Toward that end, the following is a synthesis of opinions voiced about NASA in recent years that can be filed under one of these major categories. Each perception is followed by the facts which renounce it.

PERCEPTION: NASA has no clear sense of purpose or direction, that it is in the words of one news magazine "Lost in Space".

FACT: Quite the reverse is true. Unlike past programs, such as Apollo and Skylab which were of a fixed duration, today's strategy is logical and incremental growth into the future. The intent of the Space Exploration Initiative is to make humans perma-

112

nent explorers and travelers in our solar system.

With the Space Shuttle and Space Station Freedom as the enabling technologies, the agency's long range objectives include a lunar outpost and a crewed excursion to Mars. It is significant that the lunar colony is envisioned as a permanent outpost. It will be analogous to the scientific stations operating in earth's polar regions, which depend upon resupply to sustain their operations, and support communities varying from a few to several dozen scientists. This colony, in turn, will be the technology which enables the Mars voyage. Never have NASA's objectives been so clear.

Robert Rich. Reprinted with permission.

PERCEPTION: The Space Shuttle is an ill-advised program and an illogical space transportation system.

FACT: Any spacefaring nation would characterize an optimum transportation system as being a reliable, reusable, man-rated, heavy-lift booster, carrying crews and significant cargos to and from space—the very definition of the Space Shuttle program. The designs of ESA's [European Space Agency] Hermes and the

Soviet Buran are not a coincidence—their pedigrees are of obvious origin. The Space Shuttle is a logical progression from the early manned flight programs, and is the critical capability which makes Space Station Freedom possible, as well as the objectives of the Human Exploration Initiative.

PERCEPTION: The Space Shuttle program falls far short of its promised capabilities.

FACT: This argument usually rests on the 1970s estimates of up to 60 Shuttle flights per year, which by current standards is clearly unattainable. What these critics fail to recognize is that the early number was a best-case projection, based on a fleet of six orbiters with simultaneous launch operations at Kennedy Space Center, Fla., and Vandenberg AFB, Calif. Manpower, budget and acquisition of new facilities have also not materialized as projected. While Shuttle turnaround times have not been as good as was hoped, the story is nowhere near as dismal as critics suggest.

PERCEPTION: The Space Shuttle is fragile and is crippled by technical problems.

FACT: Many critics want to apply assembly line values to the space program, as though it were comparable to the manufacture of automobiles. Such views come from the simple failure to grasp the enormity and complexity of the program. The Shuttle is a singularly unique system unlike anything else in operation, and a great deal remains to be learned about it as operations continue. Problems, such as recent difficulties with hydrogen leaks, will continue through the duration of the program and must be regarded as inevitable and the natural consequence of moving technology forward.

Freedom Will Succeed

PERCEPTION: Space Station Freedom is conceptually unsound.

FACT: This perception results from a very shallow understanding of the Preliminary Design Review (PDR) process. The PDR is the point at which the program makes a "reality check" by comparing its specifications to its capabilities. It is a given that, going in, costs, weights, man-hours, and a number of other ingredients will be prohibitively high. But this process scrubs programs back to realistic levels to ensure compatibility with program requirements. Nevertheless, preliminary figures have resulted in criticism that the program is fundamentally flawed. In fact, the results of these reviews have been expected; many minor issues to work on, a few more significant issues, but no "show stoppers." Apollo, Skylab, Space Shuttle, and all other major programs have passed through the PDR process, and Freedom is in good order by comparison. Despite funding instability, the program passed through fiscal year 1990 adjustments without an increase in total development cost. The first element launch date is unchanged and the program remains on budget. . . .

A Great Success Story

Despite all the negativism surrounding the space program, it has done great things. It has been one of the success stories of this country, and we've got to be careful not to lose that as we try to fix the current problems. Society would be much poorer materially and spiritually if we were not in space.

John Logsdon, *People Weekly*, July 30, 1990.

PERCEPTION: No market demand exists for Space Station Freedom.

FACT: The program has a large and enthusiastic constituency of domestic users and international partners. In fact, Freedom is already fully utilized in terms of available power and rack space.

Sophistication Equals Reliability

PERCEPTION: NASA prefers big, costly programs and resists the "Big Dumb Booster" concept.

FACT: "Big Dumb Booster" is a catchy phrase which suggests a heavy-lift booster system can be developed with raw power, but without high-tech and expensive avionics and electronic sophistication. Realistically, costs increase proportionally with payload capability. The argument that cost might be cut by reducing levels of sophistication, does not recognize that sophisticated systems, testing, and the redundancy which drives costs up, translate into reliability. It is also true that nothing is more costly than low reliability. No launch customer is going to want to put a very heavy (and hence, very costly) payload on an unreliable booster.

PERCEPTION: NASA has grown excessively large in budget and manpower.

FACT: By any measure—as a fraction of the federal budget, relative to Gross National Product, or by any other economic indicator—the NASA budget has declined in relative terms for the past two decades. At its peak during the Apollo program, NASA received more than four percent of the Federal Budget. During fiscal year 1990 the agency's budget was 1.6 percent of the total. The number of NASA employees peaked at close to 36,000 in fiscal year 1967. Presently the agency is smaller by approximately 1/3, with fewer than 24,000 employees.

PERCEPTION: Federal expenditures on the space program are detrimental to the national economy.

FACT: A number of independent studies show NASA programs promote economic growth with cost-to-benefit ratios as low as $6-to-$1, up to as high as $35-to-$1. Chapman Research Group Inc. contacted more than 400 companies to evaluate the impact of NASA spinoffs, and estimated cost savings of $22 billion from applied technologies. In 1988 the Nobel Prize in Economics went to an MIT [Massachusetts Institute of Technology] professor, Dr. Robert Solow, in part for his work on a mathematic model which showed how a culture's economy prospered from its investment in technological progress. NASA's expenditures do not somehow vaporize in space—they are invested in the American economy and industry.

PERCEPTION: Public support for NASA has eroded.

FACT: Every indicator of public interest in NASA is at its highest level. These include volume of public mail, numbers of visitors to the NASA field centers, Speakers Bureau requests, and educational requests from both teachers and students. Public opinion surveys consistently show approximately 80 percent approval rating for the agency, and a similar percentage believe NASA funding is adequate, or should be increased. Moreover, Congress has increased NASA's budget steadily over recent years. In a zero-growth budget environment such as is legislated under Graham-Rudman-Hollings, this is a genuine gesture of support.

It is interesting to note by comparison the numbers of media covering manned launch operations. Apollo 11 and STS-1 [first space shuttle launch] were high points with approximately 2,700 media attending each of those launches. By the 10th Shuttle flight the number dropped to 883, and Mission 61-C—the flight immediately prior to the Challenger accident—attracted 380 news media representatives. The Oct. 1990 launch of Discovery drew 554. These figures warn against confusing public interest with media interest.

NASA's Strengths Endure

The point is not that NASA is perfect. Human frailty and fallibility assure that nothing we do will ever be perfect. The Apollo program encountered difficulties, as have unmanned missions such as Viking to Mars, Voyager which toured the planets, and Magellan to Venus. Yet in each case the resiliency, innovation, and skill characteristic of NASA has persevered. Most observers outside the aerospace community can't know programs and operations in sufficient detail to appreciate this, and accordingly are vulnerable to misinformation. The only way to dispel these misconceptions is with fact.

"The most fundamental NASA problem is that every year the United States spends more to accomplish less in space."

NASA Cannot Competently Explore Space

Gregg Easterbrook

In the aftermath of the *Challenger* disaster and the flawed Hubble telescope, NASA has faced increased skepticism from the media, legislators, and the public. In the following viewpoint, Gregg Easterbrook argues that NASA can no longer effectively explore space because it is plagued by mismanagement, soaring expenses, and a lack of innovation. Easterbrook proposes that NASA be radically reformed by reducing personnel, allowing other government agencies to propose space missions, and canceling Space Station *Freedom*. Easterbrook is a contributing editor for *Newsweek* and the *Atlantic*.

As you read, consider the following questions:

1. According to Easterbrook, why does NASA resist using newer, less expensive rockets?
2. Why does the author believe that continued reliance on the shuttle threatens space exploration?
3. How do aerospace industry interest groups harm the space program, in Easterbrook's opinion?

Conventional wisdom holds that in the days of the moon race, the space program was exorbitantly expensive. Actually, measured by achievement, it was cheap. For the inflation-adjusted price of the [more than] forty space shuttle flights so far, the National Aeronautics and Space Administration could have launched 150 *Apollo* missions to the Moon. Can't believe that space used to cost less? Test-fire this statistic: during the Moon race, NASA's fanciest rocket put payload in orbit at about $1,700 per pound in current dollars. Today the comparable figure for the space shuttle is about $11,600 a pound.

Recent U.S. space history contains one lapse after another: *Challenger*, the Hubble telescope, the antenna failure that imperils the Galileo probe bound for Jupiter. NASA will reach a new low when it launches the costly GOES-NEXT weather satellite, which the agency already knows to be defective. But the most fundamental NASA problem is that every year the United States spends more to accomplish less in space.

This situation can be changed. There are many affordable, beneficial, and important things that can be done in space if only NASA truly were reformed. Here's a guide to how.

Exorbitant Launch Costs

The single greatest handicap of the space program is NASA's insistence on doing nearly everything via the space shuttle—a flying machine that is technologically impressive, visually enthralling, and fiscally nonsensical. Reusable spaceships sound cheap in theory, and someday they may be in fact. For the moment they are more expensive than the old throwaway boosters. The Saturn V, which powered *Apollo* to the Moon, cost $415 million per launch (all money figures in this viewpoint have been converted to current dollars) and put 250,000 pounds of payload in space. The shuttle now costs at least $515 million per launch and carries 48,000 pounds of cargo to space. The shuttle program was originally justified as one that would cut the cost of access to space to $400 per pound, meaning that on the key item for an efficient space program—boost cost to orbit—the shuttle is twenty-seven times more expensive than NASA predicted. The Soviet Union, with less than half the economic output of the United States, conducts on average seven times as many annual space launches mainly by using "big, dumb boosters," not max-tech hardware. The USSR's primary manned booster, the SL4, costs about $900 per pound of payload.

NASA resists ideas for new rockets because if any cheap booster were built, the shuttle's funding would be threatened. That would mean not only astronauts out of work, but also fewer jobs in the agency's manned flight centers, where there are several hundred support employees for each skywalker.

The U.S. space inventory contains some unmanned boosters that are less expensive than the shuttle, but nowhere near as cheap as they might be. All were conceived in the 1950s as military ICBMs: their designs embody assumptions from what is now space engineering's Stone Age. A combination of "dumb" design (using the least cost, not the highest-tech solution to any problem) and engineering breakthroughs has made throwaway rockets steadily cheaper since the 1950s: the powerful MX costs *less* than the old Atlas ICBM. But the civilian space program has not benefited from such advances. The United States has designed just one rocket specifically as a space booster, the Saturn. Saturns were advanced in some aspects, dumb in others: the first stage fuel was kerosene. Saturn is the only rocket built by any country to have achieved a flawless operating record. Yet NASA couldn't wait to get it out of production in the 1970s, because its continued existence would have embarrassed the shuttle.

© 1986, John Trever, *The Albuquerque Journal*. Reprinted with permission.

In 1964 NASA had 1,477 employees per space mission launched. By 1974 the figure had risen to 1,625. In 1984 there were 1,823. In 1990 there were 2,953 employees per launch. This bloat has occurred during a period when computers have revolutionized the efficiency of technical operations and knowl-

119

edge of space flight has increased. Today it ought to be elementary to launch missions with fewer people, not more. Imagine any private company trying to compete with twice as many workers per product as in 1964.

NASA Repeats Shuttle Mistakes

The 1990 Augustine Commission called a second shuttle tragedy "likely" unless most flights are replaced with throwaway boosters. There may be some causes in space worth dying for. Can releasing a data relay satellite (*Challenger*'s mission) really be one of them?

When a space shuttle fails on a mere satellite launch, not only are precious human lives lost, but billions in hardware and one-quarter of the nation's space fleet vanish too, and the entire space program goes into suspension. These things do not happen when a throwaway rocket, never expected back in the first place, malfunctions.

NASA continues to repeat the mistakes that led to the *Challenger* disaster. Despite the shuttle's vaunted new "escape mechanism," there is still no possibility astronauts could survive a repeat of the failure in the solid-rocket motor that destroyed *Challenger*, or of an uncontrolled emergency in the main engines, which have nearly miscarried on several ascents. The escape mechanism can be used only after all five motors have been safely shut off; and after the shuttle, otherwise undamaged, has been steered back to low altitude and slowed down. This covers only a small percentage of shuttle calamity scenarios.

NASA's former administrator, Richard Truly, is an ex-shuttle pilot. Unlike some past administrators, he is an able and dedicated leader. "But Truly sees his task as to protect the astronaut's franchise and increase shuttle funding on his watch," said a member of a space commission. "He can't talk about anything but shuttle, shuttle, shuttle." The shuttle hard core constantly declares itself the last line of defense between the final frontier and the end of the manned space program. The truth is that the shuttle, soaking up all funds that might go to improved new launchers, is the greatest threat to manned space exploration. If another shuttle goes down in flames, Congress may cancel future operations, leaving the United States with no space transportation whatsoever. . . .

Freedom: A Mismanaged Project

When NASA proposed the space station in 1984, the basic construction price was put at $11 billion. Today, with the station shrunk to half its former size and nearly all features deleted, NASA admits the price is $30 billion. The GAO [General Accounting Office] says $40 billion. Once the overruns set in, what will the real number be? Since agency managers clearly lied about the 1984 figures—three times the cost for half the size is not an honest mistake—Congress should assume that NASA continues to lie even about the recent inflated estimates. The more significant figure is $118 billion, the GAO estimate of what Freedom will cost once operating expenses are included. That number represents sixty years of funding for the National Science Foundation.

The station is already on its fifth design overhaul. "This is the most screwed-up project I've ever been associated with" is the line heard from every NASA and contractor station official who talks off the record. Technical worries about Freedom turn on its launch packaging and maintenance. The first U.S. space station, Skylab, went up in 1973 as a single unit, no assembly required. So that components can be launched aboard the shuttle, Freedom is being designed in small units that will require twenty-three to twenty-six shuttle flights for delivery, and in-space assembly far more complex than any ever attempted. (Saturn Vs, if they still existed, could launch Freedom in five flights.) Suppose another shuttle fails during the construction period, and launches undergo the kind of lengthy suspension that followed *Challenger*. The Freedom design can withstand one to two years of flight suspensions without falling back into the atmosphere. Downtime after *Challenger* was three years.

And because of all the component package assemblies necessary to cross-justify the shuttle, Freedom will require an unprecedented degree of maintenance. In 1990 NASA tried to suppress an internal report estimating Freedom may need 2,200 hours per year of spacewalk time for maintenance; later the estimate rose to 3,700 hours. So far American astronauts have accumulated about 400 total hours walking in space, and they found the experience profoundly exhausting. Under the current estimate, each member of Freedom's four-person crew would spend two hours per day spacewalking with wrench and hammer. In other words, the main purpose of being on the space station will be to maintain the space station. . . .

An Overpriced Launch System

Under pressure from Dan Quayle's Space Council, NASA announced it would build a new family of throwaway rockets to be called the National Launch System.

On paper the NLS sounds great—a big, dumb booster derived from existing components or new hardware that avoids complexity. For example, the NLS would have liquid-hydrogen engines only slightly less powerful than the shuttle main engines, but with a simplified internal structure involving hundreds fewer moving parts. Here's the problem: NASA says the NLS will take ten years and $12 billion to develop. In the early

1960s, when there were no CAD/CAM [computer-aided design/computer-aided manufacture] computers, the Saturn 1B, similar in features and performance to the NLS, took four years and $4.2 billion to develop. NASA appears to be employing the reverse of the old low-ball gambit, overpricing the NLS so that the agency can claim it tried to construct a relatively cheap alternative to the shuttle but could not win backing from Congress. . . .

The Research and Development Illusion

Space is widely assumed to be a "technology-driven" pursuit; Quayle has cited techno-competitiveness as a reason to build Freedom. In fact, since the shuttle designs were completed in the late 1970s, NASA has contributed little to technology. "You can debate the purpose of Star Wars, but nobody doubts that the SDI Organization, with one-third the budget of NASA, is producing more interesting new technology," said Bruce Murray, a former director of NASA's Jet Propulsion Laboratory.

The Failure of NASA

NASA, the proud agency that once bore the nation's hopes in space, has outlived its usefulness. It is an aged and faltering institution whose shortcomings have become embarrassingly evident to all but the most self-interested.

George Henry Elias, *The New York Times*, August 26, 1990.

Wander through mission control in Houston and you behold banks of flight controllers staring at screens. What do they see? Column after column of marching numbers being spat out by antiquated 1970s-style central processors. Watching for an important number is "like scanning the telephone book trying to spot a typo," says a NASA consultant familiar with the system. Modern computers could highlight any information needing attention, dropping insignificant data to the background. NASA has resisted this idea, partly because it would reduce the need for controllers.

Space R&D has declined because the bulk of NASA's appropriations now goes to shuttle subsidies. "During the old days NASA was good at doing new things, driven by technology," Murray says. "Now NASA mainly does shuttle operations, the kind of activity better left to private enterprise, while the flair for the new has been lost. And the big proposal is to bog NASA down in operations even further, by creating space station ops.". . .

Unlike some public issues, space is not critical; should GOES-NEXT or Galileo fail, life will go on. But space is a symbol of

122

government's willingness to face its own problems—if we can't fix the space program because of pork barrel and deadweight bureaucracy, how can we hope to tackle intractable problems? The U.S. space program could be cheaper, more effective, and more flexible. All that's required is someone in the White House willing to face down entrenched aerospace interest groups. Here's the plan:

What NASA Should Do

Cancel station Freedom. If a true need for a space station someday develops, one can always be built later. Use half the money saved to develop new generations of dumb boosters for cargo and small spaceplanes or SSTOs [single stage to orbit vehicles] for people, cutting the cost of access to orbit and rationalizing the U.S. launch program. Use the other half for cost-effective science, especially new robot probes not of the far heavens (fascinating, but irrelevant to daily life on Earth), but of Mars, Venus, and the Sun. Cut NASA overstaffing by a third, which alone will speed up initiatives like the environment "crash program."

Convert the shuttles to a fleet of long-duration orbiters that will be launched at a safe and affordable rate—two or three times per year—for life-science research and other experiments. Do those experiments jointly with the Soviet Union. Conduct research into ideas like nuclear propulsion that might make Moon bases and Mars exploration practical, but don't hold your breath till they actually happen. Break NASA's monopoly by allowing the Departments of Defense and Energy to propose civilian space missions; this would force NASA to compete for its budget share.

NASA officials are correct to say that humanity must learn to "live and work in space." But it's spaced-out logic to think we should be jumping directly to that purpose when we haven't even learned sensible ways to get into orbit.

"The space telescope will serve as a window on the birth of the universe. "

The Hubble Space Telescope Is a Symbol of NASA's Success

F. Duccio Macchetto

Despite flaws in the Hubble Space Telescope, many astronomers praise its discoveries and sharp photographs. In the following viewpoint, F. Duccio Macchetto argues that the Hubble Space Telescope is a great success because it provides the clearest possible images of other galaxies. Macchetto contends that the telescope will increase astronomers' knowledge of celestial mysteries, such as black holes and the age of the universe. Macchetto is head of the Science Programs Division at the Space Telescope Science Institute in Baltimore, Maryland. He was responsible for the design and implementation of the Hubble Telescope camera.

As you read, consider the following questions:

1. What important facts did Hubble critics overlook, according to Macchetto?
2. According to Macchetto, how has the Hubble flaw impeded space science?
3. Why does the author believe that earth telescopes are less effective than the space telescope?

F. Duccio Macchetto, "Points of Light." Excerpts from this article are reprinted by permission of *The Sciences* and are from the January/February 1992 issue. Individual subscriptions are $18 per year. Write to The Sciences, 2 East 63rd St., New York, NY 10021 or call 1-800-THE-NYAS.

In 1971 NASA began in earnest to lay the groundwork for what was then referred to as the Large Space Telescope. Six years later Congress approved the design of an instrument with an aperture 2.4 meters (about ninety-four inches) across. By that time the European Space Agency, of which I am a member, had become a partner in the project. In 1986 what is now known as the Hubble Space Telescope, named in memory of the American astronomer Edwin Powell Hubble, was scheduled to be launched from the space shuttle; the launch, however, was postponed several times in the late 1980s because of various technical glitches and a general slowdown in the space shuttle program following the *Challenger* accident. Finally, on April 24, 1990, the 43.5-foot-long, twelve-ton, $1.5 billion space telescope was launched and inserted into an orbit 380 nautical miles above the earth.

Blurred Images

What took place next is one of the most distressing tales of dashed expectations in all of science. Shortly after the launch a series of photographic test images was made of the bright star Iota Carinac and transmitted to earth. Astronomers, eager after years of frustration for the crystal clarity of the space-based views, saw instead a blurry image of what ought to have been a dimensionless point of light. The ensuing flurry of concern soon turned to horror: it was discovered that the shape of the primary mirror, the first station on the journey of the starlight through the telescope, was seriously flawed.

The flaw was diagnosed as a spherical aberration, a relatively common phenomenon in ground-based telescopes. But it was devastating news to the people who had spent five years building and lovingly polishing the mirror to what all assumed was a perfection unsurpassed by any earlier optical device. A subsequent investigation, led by General Lew Allen, director of the Jet Propulsion Laboratory in Pasadena, revealed that "most, if not all, of the . . . problem can be traced to a spacing error in the reflective null corrector," a device used to monitor the polishing of the mirror to its desired paraboloid shape. That disclosure sent shock waves through the astronomical community. Politicians and commentators cited the Hubble folly as a prime example of the failure of big science, a catastrophic waste of tax dollars.

Two critical facts have been obscured in all the turmoil. First, the error will be repaired, at a relatively low cost ($30 million, compared with the $200 million annually spent on the project for current operations and construction of replacement units), perhaps by the end of 1993. Second, even a flawed space telescope produces images considerably sharper than the ones from ground-based telescopes. In particular, imaging and spectroscopy in the ultraviolet wavelength region of the spectrum can be done only from space, because the earth's atmosphere—even with a depleted ozone layer—absorbs most ultraviolet radiation. Access to that band, even if somewhat degraded by spherical aberration, promises new insights into many mysteries of the cosmos: black holes, quasars and the age, nature and origin of the universe.

How Hubble Works

In principle the space telescope operates much the same way as the first known reflecting telescope, through which Isaac Newton gazed on the moons of Jupiter and the crescent of Venus in 1669. Each device is essentially a tube with mirrors placed at each end. In Newton's telescope the front end was left open to incoming light, whereas the back end was bounded by a concave mirror—the primary mirror. That mirror, only two inches across, gathered light from celestial objects and reflected it back toward the open end, where a smaller, planar mirror, the secondary, diverted the reflected light to the eyepiece. The observer could then view the image through the eyepiece.

Worth the Effort

Is the space telescope worth the time, effort, and cost needed to continue operating it, or should we just close it down and wait for the replacement parts to reach it in 3 to 4 years?

Here are some facts that might be useful to consider. What has really been lost is *sensitivity* not resolution, which is the ability to separate two or more objects from each other. This simply means that we can still resolve stars with the same accuracy as before provided they are bright enough. Nothing has changed here, and the HST can still do up to 10 times better in this regard than is possible from the ground. We have already started to resolve and understand the inner workings of a number of complex, relatively nearby objects which were just a blur from the ground.

Francesco Paresce, *The Christian Science Monitor*, October 31, 1990.

In the space telescope light also enters the open end and strikes a primary mirror—this one almost eight feet in diameter. The light is then reflected back to a convex secondary mirror about a foot across. The two mirrors are 193 inches apart, a distance that must be adjusted to within one ten thousandth of an inch if the space telescope is to remain in focus. Light hitting the secondary mirror is not, as in Newton's device, deflected to an eyepiece. (The observer's cage and eyepiece intended for a live, warm astronomer are all but extinct in the modern age of automated telescopy.) Instead the light bounces back through a hole in the dead center of the primary mirror. Having traversed that hole, the light can then be analyzed by a range of sophisticated, computer-driven instruments.

Hubble's Excellence

What lends the space telescope its distinction, however, is not its size: its primary mirror is not quite half the diameter of the huge mirror at the Mount Palomar Observatory in California. Rather it is the vastly enhanced resolution, or sharpness of image, that makes the space telescope so special. Resolution is expressed as the smallest angle between which two adjacent sources of radiation can be differentiated. The space telescope was built to discern an angular resolution of 0.05 second of arc, which is roughly equivalent to being able to spot the top of a person's head at the earth's surface from the telescope's orbital altitude of 380 nautical miles. Such a capability makes it possible to distinguish two objects that are extremely close to each other in the sky or to discern extraordinarily subtle details of a celestial object—be it the center of a distant galaxy or the surface of Saturn. The sharp resolution of the space telescope, which extends from the ultraviolet band of the spectrum, through the visible and on to the infrared, is expected to remain far superior to that of the best ground-based telescopes for many years to come: it stands as the greatest advance in resolution since the invention of the telescope almost 400 years ago.

In that context the spherical aberration detected since the launch of the telescope is particularly disappointing: the images from the space telescope are simply not as sharp as planned. The aberration comes about because on some parts of the primary mirror the shape of the surface deviates from its ideal paraboloid to that of a sphere. Because of the defect, light rays from any two objects reflected at different distances from the center of the primary mirror do not focus at the same point. Without the pinpoint focus, the space telescope gives rise to images that tend to be sharp at the center but are surrounded on the periphery by a much larger, fainter, diffuse halo. . . .

The flaw in the mirror has had a serious impact on the space telescope science program. A large number of programs have had to be delayed until the problem is corrected: among them are the Cepheid project, designed to gauge the distance scale of the universe; the search for planetary systems around nearby stars; and the observations of many of the faintest and, typically, most distant objects—particularly the ones thought to be nascent galaxies. A mission on which space shuttle astronauts will install compensating mirrors to restore the optical capability of the telescope to its advertised precision is scheduled for November 1993.

But even an imperfect space telescope can far outshine terrestrial observatories. . . .

Clarity in Space

To be sure, in the time it took to design, develop and finally launch the space telescope, enormous improvements were made in ground-based astronomy—progress that will continue. Because of some intrinsic limitations in the ground-based approach, however, there will always be a need for space telescopes. Consider the path of starlight through the earth's atmosphere. As the light enters the upper atmosphere, it is bent to varying degrees along the way, which makes the image of a star appear to shift its position in the night sky. The result in large telescopes is an image made up of myriad faint, moving speckles, which the camera records as a somewhat blurred image.

The space telescope avoids that problem as well as another one just as crucial to students of the universe. The atmosphere acts to filter out much of the light in the electromagnetic spectrum—virtually all but the narrow band of wavelengths in the realm of visible light. In this way the atmosphere cheats ground-based astronomers out of analyzing some of the most important electromagnetic radiation in the universe—the infrared, for instance, the ultraviolet and the X-ray band. Space telescopy can also eliminate the problem of "air glow": extraneous light in the atmosphere. The air glow acts like the static on a radio, making it difficult to detect the dim light signals from distant objects against the noisy glow of the background.

Significant Discoveries

The Hubble Space Telescope has surprised astronomers by taking a new head count of a stellar "baby boom" in a neighboring galaxy, turning a mere engineering test into the first proof that the orbiting observatory will still be able to make significant findings despite its flawed mirror.

Kathy Sawyer, *The Washington Post National Weekly Edition*, August 20-26, 1990.

But it is not merely its position above the atmosphere that gives the space telescope its unique capabilities; the observatory also benefits from a broad array of sophisticated scientific instruments. The faint object camera, for instance, which collects information about objects at the greatest possible distances, is made up of two complete detector systems. Optical relays inside the camera magnify the high resolution image gathered by the primary mirror. Incoming starlight, even the feeblest, is then significantly amplified by an electronic image-intensifier tube, which can precisely record the arrival time and energy of individual photons. The device creates an image in the camera more than 100,000 times brighter than the image focused by the telescope mirror.

Other instruments in the Hubble observatory include a wide-field camera, which has the largest field of view on board; a high-speed photometer, which measures fluctuations in the output of light from stars and galaxies as rapidly as 100,000 times a second; a guidance system, which accurately measures the positions of astronomical objects; and two kinds of spectrograph.

The Age of the Universe

Why all the trouble? What kinds of questions could astronomy possibly address that require such a bravura technical performance? One of the priorities of the space telescope mission is to determine the distance scale of the universe. Distances to stars and galaxies are commonly—and misleadingly—given as some precise number of light-years from the earth. But such measurements are only broad estimates. Astronomers have yet to determine precisely the distance scale of objects beyond the Milky Way; indeed, some remote galaxies may be either half as far or twice as far as their generally accepted distances from the earth.

The problem of distance is more than just a concern of celestial atlas makers. Rather it is central to most key problems in astronomy, in particular, to cosmological questions about the age and origin of the universe. In the 1920s Edwin Hubble began to chart the phenomenon known as redshift, a Doppler effect observed in the spectrum of an astronomical object, which measures how fast the object is receding from the earth. By charting the redshifts of numerous galaxies, Hubble determined that most of them are moving rapidly away from the Milky Way, and that the speed of their recession increases with distance; the magnitude of that increase (in essence, the constant of proportionality) is given by the Hubble constant. For galaxies whose recessional speed approaches the speed of light, the Hubble constant gives the size of the observable universe, and its inverse has been loosely interpreted as the approximate age of the universe. . . .

The highly sensitive Hubble Space Telescope should help gather information about many new galaxies, some of which may be in the early stages of genesis. By probing quasars, and the formation of some of the first galaxies, the space telescope will serve as a window on the birth of the universe.

"The [Hubble] flaw has become a symbol to the public of what is wrong with NASA."

The Hubble Space Telescope Proves NASA's Ineffectiveness

Eric J. Lerner

Flaws in the orbiting Hubble telescope, which were detected in June 1990, rank as one of NASA's more serious setbacks. In the following viewpoint, Eric J. Lerner argues that poor NASA decision making resulted in a misshaped mirror that eluded detection until the telescope was in orbit. Lerner asserts that NASA rejected additional performance tests on the mirror to reduce production cost and time. He maintains that NASA regularly sacrifices reliability to lower costs and meet deadlines. Lerner is the electronics editor for *Aerospace America*, a monthly aerospace magazine.

As you read, consider the following questions:

1. According to Lerner, why was NASA anxious to lower the Hubble telescope's production cost?
2. In the author's opinion, how could NASA have prevented the faulty mirror?
3. Why does Lerner believe that NASA failed to adequately investigate the Hubble error?

Adapted from "What Happened to Hubble" by Eric J. Lerner, *Aerospace America*, February 1991. Reprinted with permission.

In November 1990, a special NASA Board of Investigation chaired by then Jet Propulsion Lab Director Lew Allen published its report on the causes of the massive flaw in the Hubble Space Telescope's primary mirror. The report concluded that the flaw, a spherical aberration due to an incorrect figure or shape of the mirror, was caused by a single inaccurate measurement during construction of the instrument for checking the mirror's figure. It was not intended to determine who was to blame, and in fact the board went to some lengths to avoid doing so. However, the most prominent conclusion was that the technical team at Perkin Elmer that produced and tested the mirror operated in a "closed shop" and, without informing anyone else, dismissed three indications that the mirror was shaped wrong.

This conclusion was widely circulated by the press: "Allen panel blames the telescope's fuzzy vision on opticians who trusted their equipment more than their eyes," read a headline in the prominent journal *Science*. The article faulted NASA and Perkin Elmer management for sloppy oversight, but placed the main fault at the bottom of the totem pole.

NASA's Wrong Conclusion

Though widely circulated and comforting to NASA management, this explanation is fundamentally wrong. A special investigation by *Aerospace America*, which included interviews with a dozen participants in the Hubble project at Perkin Elmer and NASA, and a review of the Allen commission's reports and the history of the Hubble Telescope project, turned up overwhelming evidence of a different problem.

Technical errors were certainly made, but the main reason they were not caught was a consistent NASA policy to eliminate cross-checking of vital measurements as "unnecessary." As Dan Johnston, NASA's engineering representative on the mirror project, put it, "The philosophy was: if you made the measurement correctly once, you don't need to make it a second time." This systematic elimination of all redundancy in the testing of the mirror stemmed from the overwhelming imperative, originating at the highest levels of NASA, to reduce the time and cost of the project to an absolute minimum, an imperative ultimately driven by the ever-growing financial needs of the troubled Space Shuttle program, which similarly squeezed every other NASA project in the '80s. "Doing each measurement once was the most cost-effective way," recalls Project Scientist Robert O'Dell, "and cost was the name of the game on this project."

Again and again, additional tests that would have exposed the flaw were proposed and turned down. Indeed, Program Director Fred Speer at NASA-Marshall repeatedly attempted to eliminate tests already in the program.

Given this policy of ruling out cross-checking, once a mistake was made in a measurement, it would remain undiscovered. The three clues that something was wrong with the mirror were not kept secret. None was pursued, however, despite proposals from the mirror polishing team, because that would have involved rechecking measurements already made, which was forbidden.

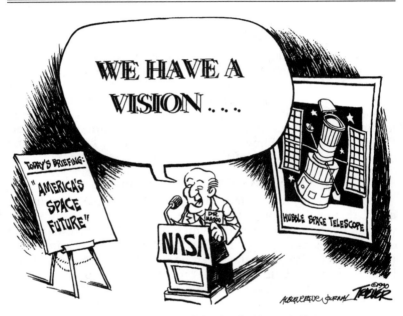

© 1990, John Trever, *The Albuquerque Journal*. Reprinted with permission.

This primary problem of doing the project by the fastest and cheapest methods was compounded by an organizational problem. No one in or out of NASA had the personal responsibility for the telescope mirrors that the principal investigators (PIs) had for the five instruments that made up the rest of Hubble's complement. These PIs, based in universities outside NASA's administrative chain and armed with the authority that came from designing the instruments, acted as advocates, defending the instruments against NASA corner-cutting. In contrast, the telescope mirrors were orphans.

The causes of the Hubble Telescope flaw are important not only because the telescope is a major scientific project, but because the flaw has become a symbol to the public of what is wrong with NASA, the American space effort, and big science generally. Scientists and lay people alike wonder how a mirror reputed to be the most precise in the world, accurate to 1/64th of a wavelength of light, could turn out to be in error by 30 times as much; how an error that could have been detected with a pinhole light source and knife edge, the Foucault test used by generations of amateur scope makers, could have slipped through NASA quality control; how such an error could remain undiscovered when two test instruments indicated something was seriously wrong. . . .

Extra Testing Rejected

Perkin Elmer's scientists and engineers, who were experienced in instrument making but not in the construction of such a large telescope, were confident that they could get it right the first time, and they were not alone. NASA-Marshall's Charles Jones, who had the primary technical responsibility for overseeing the telescope's optical fabrication, told *Aerospace America*, "We thought that it was a reasonable risk. I published a paper on the approach in a major journal, *Optical Engineering*, highlighting the fact we weren't doing an end-to-end. There were no letters to the editors saying we were foolish."

Yet, before polishing started, proposals were made to put cross-checking into the program. NASA had assigned Eastman Kodak to build a back-up mirror in case anything happened to Perkin Elmer's. Kodak's team, according to [retired vice president] Charles Spoelhof, recommended that when the mirrors were completed they be tested by a common set of instruments. That test would certainly have revealed the Perkin Elmer flaw. Within NASA, Dan Johnston, who was the on-the-spot engineering representative to the polishing team, also wanted more tests. "I suggested additional tests on more than one occasion, including a full-up system test. I also proposed using a reflective and refractive null simultaneously on the Perkin Elmer mirror," recalls Johnston. Even Perkin Elmer proposed some limited additional tests. "We had in our proposals an end-to-end test on the central 20 inches of aperture," says John D. Rehnberg, vice president of Perkin Elmer's Space Science Divisions.

These proposals for new tests "simply didn't have a snowball's chance," O'Dell says. By the time polishing began in the summer of 1980, the Shuttle program was a billion dollars over budget and years behind schedule. NASA had to seek additional funding from Congress in midyear. Edward Boland, chairman of the House Appropriations subcommittee that dealt with NASA, suggested part of the money needed could be cut out of the Space Telescope program. But the Space Telescope's needs were growing rather than shrinking. It too was greatly exceeding its budget, which had been rather optimistically estimated to get the project approved. Fred Speer, a manager with a record of completing projects on budget and on schedule, was brought in as project manager in February 1980. [NASA] relayed a stern message to Speer: Stay within budget and on schedule or the project might be canceled.

One of NASA's Many Problems

Hubble may justifiably be seen as merely the tip of NASA's iceberg. The agency has received a great deal of criticism: for a space shuttle fleet that costs too much and underperforms, for increases in the time and expenditures required for major projects. But it has been hard for critics to present well-defined critiques rather than mere complaints. After all, no other organization is flying a shuttle fleet or launching the largest of spacecraft. This lack of alternatives to NASA precludes standards by which the agency's performance might be judged.

The Hubble problem is different. It involves a straightforward technical issue that other outfits have successfully dealt with, but which NASA blew.

T.A. Heppenheimer, *Reason*, October 1990.

Just as polishing began, Speer proposed saving $5 million by cutting testing. O'Dell and others successfully resisted cutting scheduled tests, but they made no headway on getting more tests. "I was asked if a new test, a cross-check, was mandatory, and I had to say no each time," recalls Johnston. Rehnberg agrees: "Unless a test was for a specific problem that was clearly and unequivocally identified and had to be fixed, NASA never supported it.". . .

Report Errors and Secrecy

Given the embarrassing nature of the Hubble error and the chain of decisions reaching right to the top, it is not surprising that NASA has been reluctant to probe deeply into responsibility. The Allen commission, according to Allen himself and two other commissioners interviewed, was instructed by NASA officials not to allocate responsibility but merely to analyze what technically caused the spherical aberration. Its report omits names of all participants in the project, and members declined to provide them when interviewed. The delicate task of placing responsibility has been assigned to NASA's own inspector general.

Nor has NASA been eager to have others investigating the causes of the Hubble problem. Upon the instruction of NASA's general counsel, the board of inquiry declined to release summaries of the testimony before it. NASA failed to respond to a Freedom of Information Act request from *Aerospace America* for this information.

Such reticence allowed factual errors to creep into the Allen report. Some, such as the misidentification of the organization at Perkin Elmer responsible for the mirror, are minor. It was the Electro-Optical Div., not the Optical Operations Div. But some, such as the assertion that the mirror team ran a closed shop, are more significant and are clearly contradicted by the routine presence of NASA personnel, like Charles Jones, during testing of the mirror. The net effect has been to obscure the role of NASA policies in the failure to detect the error and to implicitly shift all blame to technical personnel who actually mismeasured the RNC [reflective null corrector].

Sacrificing Safety

Such self-absolution would be unfortunate for NASA. The basic problems that caused the Hubble foul-up have occurred over and over. Costs of a project are underestimated, and as the unrealistic budget is surpassed, increasing risks of failure are accepted to remain within schedule and budget. This underlying chain of events led to both the Challenger disaster and the Hubble flaw. Until NASA stops trading higher risks for lower costs, tomorrow's headlines will again be asking, "Why?"

Evaluating Sources of Information

When historians study and interpret past events, they use two kinds of sources: primary and secondary. Primary sources are eyewitness accounts. For example, a National Space Council report formulating policies for future space launches would be a primary source. A newspaper article that paraphrases the council's report would be a secondary source. Primary and secondary sources may be decades or even hundreds of years old, and often historians find that the sources offer conflicting and contradictory information. To fully evaluate documents and assess their accuracy, historians analyze the credibility of the documents' authors and, in the case of secondary sources, analyze the credibility of the information the authors used.

Historians are not the only people who encounter conflicting information, however. Anyone who reads a daily newspaper, watches television, or just talks to different people will encounter many different views. Writers and speakers use sources of information to support their own statements. Thus, critical thinkers, just like historians, must question the writer's or speaker's sources of information as well as the writer or speaker.

While there are many criteria that can be applied to assess the accuracy of a primary or secondary source, for this activity you will be asked to apply three. For each source listed on the following page, ask yourself the following questions: First, did the person actually see or participate in the event he or she is reporting? This will help you determine the credibility of the information —an eyewitness to an event is an extremely valuable source. Second, does the person have a vested interest in the report? Assessing the person's social status, professional affiliations, nationality, and religious or political beliefs will be helpful in considering this question. By evaluating this you will be able to determine how objective the person's report may be. Third, how qualified is the author to be making the statements he or she is making? Consider what the person's profession is and how he or she might know about the event. Someone who has spent years being involved with or studying the issue may be able to offer more information than someone who simply is offering an uneducated opinion; for example, a politician or layperson.

136

Keeping the above criteria in mind, imagine you are writing a paper about NASA's role in space exploration. You decide to cite an equal number of primary and secondary sources. Listed below are several sources that may be useful for your research. *Place a P next to those descriptions you believe are primary sources. Place an S next to those descriptions you believe are secondary sources.* Next, based on the above criteria, *rank the primary sources, assigning the number (1) to what appears to be the most valuable, (2) to the source likely to be the second-most valuable, and so on, until all the primary sources are ranked. Then rank the secondary sources, again using the above criteria.*

P or S		Rank in Importance
_____	1. A speech by NASA's administrator concerning the future of America's civil space program.	_____
_____	2. A *Time* magazine article reporting the recommendations of a space program advisory board.	_____
_____	3. A Federation of American Scientists study of planned costs and actual costs of NASA space projects.	_____
_____	4. A *National Geographic* report of space shuttle missions to rescue failed satellites.	_____
_____	5. An article by a *Washington Post* reporter describing U.S. Senate debates about space station funding.	_____
_____	6 A NASA investigative report determining the causes of the Hubble Space Telescope flaw.	_____
_____	7. A book review of *Space Enterprise: Beyond NASA.*	_____
_____	8. A magazine editorial criticizing bureaucratic problems within NASA.	_____
_____	9. An economist's comparison between government and private sector space exploration activities.	_____
_____	10. Congressional testimony by a space shuttle booster rocket engineer describing decisions that led to the space shuttle *Challenger* explosion.	_____
_____	11. A magazine article paraphrasing the opinions of several NASA managers and astronauts.	_____
_____	12. A television report describing experts' testimony before a U.S. Senate committee on NASA funding.	_____

Periodical Bibliography

The following articles have been selected to supplement the diverse views presented in this chapter.

David Bjerklie
"Roots of the Hubble's Troubles," *Time*, December 10, 1990.

William J. Broad
"Can NASA Make Space Seem Worth the Price?" *The New York Times*, May 26, 1991.

William J. Broad
"Panel Finds Error by Manufacturer of Space Telescope," *The New York Times*, August 10, 1990.

Leonard David
"NASA: Countdown to the Future?" *The World & I*, July 1991.

Amitai Etzioni
"NASA Out of Orbit," *The New Leader*, April 16, 1990.

Timothy Ferris
"Ground NASA and Start Again," *The New York Times*, March 16, 1992.

Howell Heflin
"Keep the Dream Alive," *Ad Astra*, November 1990.

Edward L. Hudgins
"America's Space Policy: Countdown to Major Reforms," The Heritage Foundation *Backgrounder*, April 25, 1991. Available from 214 Massachusetts Ave. NE, Washington, DC 20002-4999.

Leon Jaroff
"Spinning Out of Orbit," *Time*, August 6, 1990.

Jeffrey Kluger
"Lost in Space," *Discover*, January 1991.

Linda Kramer
"Despite Hubble and Trouble, an Expert Says NASA Is Worth It," *People*, July 30, 1990.

Tony Reichhardt
"Is NASA Dead?" *Final Frontier*, September/October 1990. Available from PO Box 534, Mt. Morris, IL 61054-7852.

Richard Talcott
"Hubble Delivers," *Astronomy*, May 1991.

Richard Truly
"The Exploration of Space," *Vital Speeches of the Day*, February 15, 1991.

Mark Washburn
"Do We Really Need NASA?" *Sky and Telescope*, September 1991.

Tom Wicker
"Beyond Murphy's Law," *The New York Times*, July 19, 1990.

Should Space Be Used for Warfare?

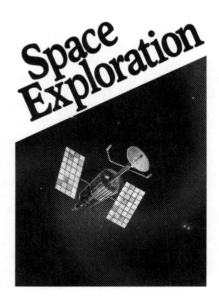

Space Exploration

Chapter Preface

During the 1991 Persian Gulf War, American military satellites in outer space detected launches of Iraqi Scud missiles aimed at Israel and Saudi Arabia. Using this data, U.S. and allied forces fired Patriot missiles to intercept most of the Scuds before they reached their targets. Largely due to these effective defenses, one of Iraq's most dangerous threats failed. For many Americans, the Patriots also renewed a sense of pride and accomplishment in the defense industry that had not been felt since World War II.

The research that led to these successful weapons is highly controversial. It is part of the Strategic Defense Initiative, which includes plans to build laser space weapons and orbiting interceptors designed to detect and destroy missiles. The plan has been variously labeled as an expensive folly by critics and the answer to the nuclear threat by advocates. After the successes of the Persian Gulf War, defense experts had proof that these weapons might succeed against future military threats. Detractors found it more difficult than ever to criticize these weapons.

Yet debate continues to surround this system. After all, the Patriots did fail to intercept *every* Scud, critics point out. If a similar weapons system failed to destroy a nuclear missile, the damage would be horrendous. The answer to the nuclear threat is not expensive space weaponry, but continued international negotiation and cooperation, many argue.

The authors in the following chapter present compelling arguments on both sides of the space weapons question. Whether space will become the next military frontier is a question that still lies in America's future.

"Space-based weapons will become increasingly important as Third World states acquire missiles of greater range."

The U.S. Needs Space Weapons to Counter Global Military Threats

Thomas G. Mahnken

While the end of the Cold War has diminished the threat of a missile attack against the United States, many Third World nations now seek longer-range ballistic missiles, even nuclear weapons. In the following viewpoint, Thomas G. Mahnken argues that the United States needs space weapons to protect itself from such potentially hostile nations. Mahnken maintains that the United States must also develop space weapons to counter nations intent on using satellite military data against the United States or its allies. Mahnken is a national security analyst for SRS Technologies, a defense contractor located in Arlington, Virginia.

As you read, consider the following questions:

1. According to Mahnken, how could hostile nations use improved missiles to threaten U.S. influence abroad?
2. How can satellite launch capabilities be used for military purposes, in Mahnken's opinion?
3. Why does the author believe that space weapons will improve ground-based missile defenses?

Adapted from "Why Third World Space Systems Matter" by Thomas G. Mahnken, *Orbis: A Journal of World Affairs*, published by the Foreign Policy Research Institute, Fall 1991. Reprinted with permission.

The 1991 war in the Persian Gulf illustrated vividly the impact of the developing world's acquisition of advanced military technology. Saddam Husayn's ballistic missiles, chemical and biological weapons, and erstwhile nuclear program are a microcosm of the problem. However, the growth of Third World military capabilities is not limited to weapons of mass destruction. Another area that will play an increasingly important role in Third World military development is the use of space for military purposes.

Military space systems played a prominent part in the U.S. victory over Iraq, as they have in previous operations. Now such capabilities are spreading to a significant number of states in the developing world, which see high-technology weapons as necessary instruments of national policy. Currently, only seven states—the United States, the Soviet Union, China, France, Japan, India, and Israel—possess an indigenous space-launch capability, but a number of others have had satellites launched by states with such a capability.

Security Implications

The use of space by a widening circle of states has a number of implications for U.S. national security policy. Firstly, the development of Third World missile and space programs is forging a new set of geopolitical relationships over which the United States has little influence. Secondly, the spread of this technology is already eroding the political leverage possessed by the original space powers, a leverage based on their ability to control the flow of satellite information and space technology to client states. Thirdly, the spread of this technology is likely to be uneven and could lead to regional instabilities, especially if a state were to perceive and attempt to exploit a transitory military advantage in its space capabilities. Lastly, the acquisition of missile and space capabilities by rising regional powers will significantly alter the circumstances U.S. forces face when operating abroad. Collectively, these trends suggest Washington must begin formulating policies and acquiring capabilities that will allow it to retain its influence in a rapidly changing strategic environment. . . .

Ballistic Missiles

The experience of the Gulf war has made Americans aware of ballistic missile proliferation. During the forty-one days of Operation Desert Storm, Iraq launched eighty-one modified Scud missiles against the U.S.-led coalition and Israel: thirty-nine against Israel and forty-two against Saudi Arabia. While most were ineffective, a single strike on a U.S. barracks near Dhahran killed twenty-eight U.S. soldiers and wounded ninety-eight. While chemical warheads were not employed, the spectre of their use was a constant source of unease. Earlier, Iraq had used its ballistic missiles to much more devastating effect against Iran in the "War of the Cities." From February to April 1988, some 190 Al-Husayn (modified Scud) missiles were fired at Iranian cities, resulting in approximately two thousand Iranian casualties, evacuation of half the population of Tehran, and a severe disruption in the Iranian war economy.

Chuck Asay, by permission of the *Colorado Springs Gazette Telegraph.*

The then-director of the Central Intelligence Agency, William Webster, has testified that by the end of the decade, fifteen states in the developing world will possess ballistic missiles, and six of them will have intermediate range ballistic missiles. While the missiles employed by Iraq were highly inaccurate and carried relatively small high-explosive warheads, it is likely that Third World states will employ more sophisticated systems in future conflicts: missiles with greater range, higher accuracy, and more destructive warheads. According to Rear Admiral Thomas Brooks, director of the Office of Naval Intelligence, ten of the states developing a ballistic missile capability possess or are developing chemical weapons, and Webster has testified that by the end of the decade, eight of the countries seeking ballistic missiles may have nuclear weapons.

As indigenous ballistic missile technology advances in the Third World, the ability of the United States and other advanced industrialized countries to control proliferation diminishes. Furthermore, civilian space programs may conceal ballistic missile programs; any satellite launch vehicle (SLV) can be used as a surface-to-surface missile with the addition of a guidance and control package.

Ballistic missiles might allow regional powers to alter the scope, and potentially the intensity, of a local conflict in ways unfavorable to the United States. On a tactical level, ballistic missiles allow states to strike distant targets quickly, with little warning, and with a high probability of penetration. On an operational level, the possession of ballistic missiles by hostile powers threatens to constrain U.S. influence within a theater, and may jeopardize the ability of the United States to carry out operations in aid of allies. On a strategic level, the possession of such systems by regional powers—in combination with their growing economic, military, and political influence—provides grounds for challenging the leadership of the United States on regional issues. . . .

Iraqi Space Interests

A significant number of states will acquire in the next twenty years indigenous capabilities to use space to support and enhance their military operations.

Before its invasion of Kuwait, Iraq was on its way to establishing the infrastructure for a space program with military implications. On December 5, 1989, in a move that surprised many observers, Iraq tested the three-stage Al-'Abid space launch vehicle from the Al-Anbar Space Research Base west of Baghdad. Hussein Kamil, minister of industry and military industrialization, announced that the rocket would be used to place satellites in orbit. Iraq also reportedly signed a contract with the Brazilian National Institute for Space Research (INPE) and the Brazilian aeronautics manufacturer EMBRAER for construction of a military reconnaissance satellite with a 20-meter resolution, with equipment supplied by China and France. According to press reports, the deal was cancelled by the Brazilian foreign ministry in April 1989 because Brasilia, under pressure for its international arms sales, was sensitive to the satellite's military value. Given the demise of the Brazilian satellite deal, the continuing status of Iraq as an international pariah, and the devastation of Iraq's domestic infrastructure during Operation Desert Storm, the future of the Iraqi space program is at best uncertain. . . .

The acquisition of ballistic missiles, space assets, and possibly antisatellite weapons by regional powers means that they are acquiring a military capability that will allow them to exert more control over future conflicts, for example, by altering the geographic scope of a conflict through ballistic missile strikes against targets in or beyond the theater, and by deploying forces in response to accurate and timely satellite intelligence.

Zone of Protection

Let us assume that space-based Brilliant Pebbles [missile interceptors], once their development is completed, are deployed by the United States in orbits inclined at roughly 40 degrees relative to the equator. This orbital parameter would hold the BPs in a band overlooking the Earth beneath, between 40 degrees north and south latitudes. In this way, they would be positioned to engage missiles launched from locations as far north as Iraq and North Korea, and as far south as South Africa and Argentina. This 80-degree swath of mid-latitudes encompasses the vast majority, if not all, of the third world nations currently on the list of possessors, producers, and aspirants of missile capabilities and warheads of mass destruction.

John L. Piotrowski, *Global Affairs*, Spring 1991.

The acquisition of military space capabilities by regional powers could provide a local power with a critical advantage over U.S. or allied military forces in a future regional conflict. While regional powers—owing to their limited objectives—will always be able to concentrate forces in the theater of operations, the United States may not be able to achieve such a concentration, owing to its global commitments. Regional powers will also be able to reduce the United States's ability to achieve tactical—or, for that matter, strategic—surprise, owing to their ability to conduct reconnaissance from air and space. For example, the grand deception carried out by coalition forces in the Persian Gulf war—when 150,000 U.S., British, and French troops were moved one hundred miles through the desert to outflank Iraqi defensive positions—would have been greatly complicated, if not made impossible, had Iraq possessed timely data from observation satellites.

Satellite Data

Further, the availability of multiple sources of high-quality commercial observation data holds out the possibility that a regional power would be able to monitor not only activities in the region, but U.S. military staging areas in the United States and abroad. Third World powers may also acquire the ability to interrupt the flow of satellite data to the United States through

ASAT [antisatellite] attacks or jamming. While such capabilities are likely to be rudimentary, the U.S. satellite capability in distant theaters is also likely to be limited, and it might be difficult for the United States to deploy new satellites over the theater in a timely manner.

The possession of increased strategic reach by regional powers will also allow them to attack U.S. forces at a greater distance. As the Persian Gulf war showed, ballistic missiles may be targeted at U.S. or allied support bases in the theater, although their use against naval forces would be difficult without near-real-time targeting support. The possession of accurate satellite navigation and command-and-control assets, in combination with airborne early warning and maritime search forces, would also simplify the task of directing air and sea forces to meet U.S. forces.

Threats from Belligerent Nations

Regional powers may also be able to obtain satellite information from commercial suppliers of data, blurring the distinction between belligerent and neutral states. Belligerents in regional conflicts have traditionally received important intelligence information from the major powers. Speculating on what will happen once other states such as China, Brazil, India, and Japan begin to supply such data to states that might become belligerents is worthwhile. Would China supply Islamabad with satellite information to counterbalance New Delhi's indigenous capability? Would Brazil supply one of its largest arms customers, Libya, with satellite data in the event of a conflict with the United States?

If the United States hopes to retain military flexibility as space capabilities spread throughout the developing world, it must acquire forces that allow it to counter attacks on its space systems. At the same time, it must have the capacity to limit an opponent's flexibility and freedom of action.

Firstly, the United States will need the ability to monitor activities both on the earth and in space. Traditionally, the U.S. sensor network has been oriented toward gathering information on the activities of Soviet conventional, nuclear, and space forces. The acquisition of ballistic missile and space capabilities by others has produced a growing need to monitor activities by these states. U.S. space assets have proved invaluable in recent operations ranging from naval deployments in the Persian Gulf and the liberation of Panama to operations Desert Storm and Desert Shield. To quote Lieutenant General Thomas Moorman, commander of the U.S. Air Force Space Command:

> In a multipolar world, a world of diverse threats, our demands
> for global information will be much more intense. The forces

146

we deploy will require the traditional kinds of support. . . .
We'll have fewer places overseas from which to collect that
kind of information. But space systems will be on the scene,
already there, everywhere, to provide it.

For instance, U.S. satellites reportedly provided decision makers
in Washington with important information in the days leading up
to the Iraqi invasion of Kuwait, monitoring Iraqi air-defense
radars and tracking the build-up and movement of Iraqi ground
forces. The U.S. was reportedly able to use a wide variety of communications and reconnaissance/surveillance satellites in the region at the time of the invasion.

A Global Defense System

Secondly, the United States will need the ability to defend
both U.S. forces abroad and allies from ballistic missiles. The
Patriot antitactical ballistic missile was an invaluable asset in
the war with Iraq, intercepting the vast majority of Iraqi missiles within its range. However, the war has also shown the limitations of Patriot. A point-defense system, it is unable to protect
wide areas of territory. Hence several Scuds escaped intercep-
tion, one landing on an apartment building in Tel Aviv. What
will be needed is a system to provide Global Protection Against
Limited Strikes: a layered ground- and space-based system that
offers multiple opportunities to engage incoming missiles, reducing chances of failure. Key to such a system will be ground-
based theater missile defenses such as the Patriot, the U.S.-
Israeli Arrow interceptor, or the U.S. Theater High-Altitude
Area Defense system. Space-based sensors being developed as
part of the Strategic Defense Initiative will be crucial to early
warning of missile attacks, detecting launches, and assessing attacks. Further, as the Iraqi invasion of Kuwait attests, future
crises may develop literally overnight. Space-based interceptors
such as Brilliant Pebbles, constantly in orbit, are able to provide
global presence and round-the-clock ability to intercept long-
range ballistic missiles. Such space-based weapons will become
increasingly important as Third World states acquire missiles of
greater range.

In addition, as regional powers acquire the ability to use space
assets to support their military operations and threaten U.S. and
allied reconnaissance satellites, the United States will require
the means to prevent hostile states from controlling space, or
key locations in space. The U.S. Army has formulated plans to
deploy a ground-based ASAT on U.S. territory. Nonetheless, in
the current strict budget climate, the system faces a precarious
future. If the United States fails to deploy an antisatellite system, it may be forced to attack satellite ground links to deny an
opponent access to space data. Such an operation will become
all the more difficult as portable ground stations spread in de-

veloping states. Another option involves increasing the capability of U.S. satellites to defend themselves against enemy ASATs through maneuvering, protection of key components, and redundancy of vital functions.

The Persian Gulf war is a harbinger of the complexities the United States will face as it heads towards the twenty-first century. The ability of developing states to construct and maintain ballistic missiles, space assets, and ASAT weapons will give such states increased control over their regional environment. Conversely, it will reduce U.S. leverage over regional affairs. By the beginning of the twenty-first century, a number of states will likely possess a multidimensional military aerospace capability, which until now has been the sole province of a handful of developed states.

Thus, U.S. and allied forces deployed abroad will face a more difficult and complex operational environment than during the past four decades. The full realization of this situation is some years off, but a number of policy and acquisition decisions need to be made soon. If the United States hopes to retain its ability to support allies globally and enforce international stability as it has throughout the postwar era, it must obtain the capabilities that will allow it to maintain its advantages.

*"The end of the Cold War has left even Star
Wars II lacking a credible mission."*

The U.S. Does Not Need Space Weapons to Counter Global Military Threats

Kosta Tsipis

In the following viewpoint, nuclear physicist Kosta Tsipis argues that the deployment of space weapons is unnecessary because the United States faces virtually no threat from foreign missile attacks. Tsipis asserts that the end of the Cold War has eliminated the threat of attack from the former Soviet Union and China. He also maintains that Third World nations pose no threat because they lack the missile capabilities to reach the United States and would be deterred from doing so by the threat of prompt U.S. retribution. Tsipis is director of the Program in Science and Technology for International Security at the Massachusetts Institute of Technology in Cambridge.

As you read, consider the following questions:

1. Why do current arms control strategies reduce the need for missile defenses, according to Tsipis?
2. In the author's opinion, why would space weapons be useless against a terrorist nuclear attack?
3. According to Tsipis, how could the danger of an accidental missile launch be eliminated?

Adapted from "A Weapon Without a Purpose" by Kosta Tsipis, *Technology Review*, November/December 1991. Reprinted with permission from Technology Review, © 1991.

On May 26, 1972, Richard Nixon and Leonid Brezhnev signed the ABM Treaty, which banned both countries from implementing nationwide ballistic missile defenses. . . . In 1978 ABM enthusiasts attempted to advance an ABM system called LoADS (Low Altitude Defense System), this time to protect MX missiles that would be shuttling among 24 shelters arranged around a racetrack-shaped roadway planned for the deserts of Utah and Nevada. When the people of these states, with support from Nevada's usually pro-military Republican Senator Paul Laxalt, rejected this multiple-shelter basing mode, ABM advocates realized that ground-based ABM systems would probably never garner the necessary congressional support. While the Pentagon's budget soared in the first years of the Reagan administration, funding for the defensive systems remained low.

Space was the obvious alternative for basing antiballistic-missile defenses. First, Sen. Malcolm Wallop championed space-based lasers, but the Pentagon's Defense Science Board rejected the idea in 1981. Next, in 1982, came "High Frontier," a proposal to orbit thousands of rockets that could be fired against Soviet ballistic missiles outside the atmosphere. When an Air Force study found that idea technically flawed, it too quickly dissolved as a serious option. As John Gardiner, the Pentagon's director of defensive systems, told the Senate Armed Services Committee in 1983, "The entire High Frontier proposal is technically unsound. It suffices to mention that it requires the kill rockets that attack the Soviet ICBM to be fired against it 50 seconds *before* the target ICBM was launched."

Reagan's Star Wars

Within months, however, on March 23, 1983, President Reagan proposed his "Star Wars" program to render nuclear-tipped ballistic missiles "impotent and obsolete," thereby protecting the nation from Soviet nuclear weapons "just as a roof protects a family from rain." Politically and psychologically, it was ingenious. As documents of that period indicate, the proposal was intended to co-opt the anti-nuclear message of the arms-control community and assuage the fears of nuclear war that had prompted over half a million people to demonstrate against the Reagan administration nine months earlier in New York City. Moreover, its exotic, space-borne character appealed to the science-fiction-fed younger generation, while some voters were heartened to see the government summon U.S. technological ingenuity to protect them from the Soviet nuclear threat. Finally, Star Wars countered the sense of interdependence with the Soviet Union that mutual assured destruction and nuclear deterrence imposed on the United States.

The trouble was that Star Wars I, Reagan's original proposal, couldn't work and was unfathomably expensive. Two years before the president's speech, technical studies performed by the MIT Program in Science and Technology for International Security had shown that the laws of nature make space-based population defenses impossible. Laser and neutral-particle beams would lack the necessary lethality to destroy rising ICBMs many hundreds of miles away, and inexpensive counter-measures could blunt their effectiveness. Study after study since then has reached the same conclusion: it is impossible to protect civilians against an opponent like the Soviet Union. Even the most sophisticated Star Wars defenses would be vul-nerable to countermeasures that could be implemented at one-tenth the cost of the system itself.

History began repeating itself. Once it became clear that pro-tecting cities was impossible, the Strategic Defense Initiative Organization (SDIO), the Pentagon agency created to run the Star Wars project, shifted ground. Star Wars II directed essen-tially the same technology to protecting missile silos, thereby supposedly enhancing deterrence, an admission that the pre-Reagan doctrine of mutual assured destruction was still the only viable way to protect the United States from nuclear attack.

However, the end of the Cold War has left even Star Wars II lacking a credible mission, as the probability of a deliberate nu-clear attack against the United States by the Soviet Union or China is dwindling to that of attack by France or England. Thus SDIO has once again shifted the mission and name of the system originally proposed meant to protect cities. The new name is "Global Protection Against Limited Strikes" (GPALS), a space-based system that SDIO claims can deal with some post-Cold War threats but *not* with most short-range ballistic missiles like the Scud.

The events of the Gulf War added a politically resonant mis-sion for ballistic-missile defenses. But does the threat from short-range tactical ballistic missiles, or any other threat for that matter, make an ABM system militarily relevant?

Post-Cold War Threats

For at least the decade ahead, the United States might con-ceivably face any of five ballistic-missile threats.

• A current nuclear power could deliberately launch a nuclear attack.

• A near-nuclear nation—for example, India, Pakistan, or South Africa—could acquire both nuclear warheads and the long-range ballistic missiles to deliver them and launch a delib-erate nuclear attack.

• A terrorist group might acquire one or more nuclear explo-sives and use them against a U.S. city.

• An enemy might launch tactical ballistic missiles such as the Iraqi Scuds, either with nuclear or conventional warheads, against U.S. military bases or forces overseas.

• An unauthorized or accidental nuclear attack might emanate from the Soviet Union, China, or any other nuclear nation with ballistic missiles.

Scare Tactic

In an attempt to drum up support for Star Wars deployments, the Administration and the Pentagon have been conjuring up fears that Third World nations will soon have the capability to launch ballistic missile attacks against U.S. soil. The facts, however, hardly bear out these assertions.

While some Third World nations are pursuing ballistic missile programs, *the missiles in current and foreseeable inventories will not threaten the U.S.* The countries with the technology and the resources needed for long-range missiles are industrialized nations—Japan, Germany, and other allies of the United States, including Israel.

Center for Defense Information, *The Defense Monitor*, vol. XX, no. 5, 1991.

To begin with, not only is a deliberate Soviet or Chinese attack highly improbable now that the Cold War has ended, but the United States already has the proven means to prevent it. Deterrence worked through the darkest days of the Cold War; it will continue to work in the future. That implies the need for the United States to retain a secure retaliatory force of a few hundred nuclear weapons, but at the same time it obviates any need for ballistic-missile defenses. What's more, as the United States and the Soviet Union reduce their arsenals of multi-warhead land-based missiles, the need to enhance deterrence with an ABM defense for missile silos evaporates.

The same deterring mechanism that has kept Soviet and Chinese communists at bay can restrain newcomers to the nuclear club. No political leader would invite the obliteration of his or her country by attempting to destroy a U.S. city. Self-preservation and self-interest, the sturdy underpinnings of deterrence, are concerns that no national leader can ignore.

Still, Iraq's use of Scud missiles against Saudi Arabia and Israel has prompted some to assert that deterrence may not always dissuade new nuclear powers. For example, Rep. Les Aspin observed that "Saddam Hussein wasn't deterred. He faced virtual destruction of his nation, yet he still used ballistic missiles against U.S. forces and our friends." The flaw in this argument is that deterrence applies only to weapons of mass destruction, not to conventional weapons, whether delivered by plane or missile. Directing a Scud against Tel Aviv to provoke

Israel into war has little in common with a nuclear-armed missile aimed at New York—the threat we are concerned with here—or even Tel Aviv. In fact, Saddam Hussein *was* deterred from using weapons of mass destruction. Although Iraq possesses chemical weapons, the assurance of retaliation in kind apparently stopped it from wielding them either against U.S. troops in the battlefield or our allies' urban centers in the region.

No Defense Against Nukes

An ABM system, no matter how effective, can't counter the third threat, a terrorist nuclear attack, because the nuclear weapons wouldn't be delivered by a ballistic missile. A terrorist group that procured or manufactured a nuclear explosive would attempt to carry it clandestinely into the United States by means—such as boat, plane, or truck—that no antiballistic missile could intercept. Furthermore, even if a stateless terrorist group had a ballistic missile, it would be highly unlikely to use it against the United States because the attack could be traced to the country of launch, which then would face nuclear retaliation. Deterrence would apply in this case as well.

We come to the fourth threat—tactical short-range ballistic missiles with ranges up to 1,000 kilometers wielded against U.S. allies around the world or U.S. military forces abroad. This is a two-part question because the threat from such missiles carrying conventional warheads is fundamentally different from that of nuclear-tipped missiles.

Tactical ballistic missiles with conventional warheads are not accurate enough to be militarily significant. Even with sophisticated and expensive guidance systems, a tactical ballistic missile with a 1,000 kilometer range can only come within 60 to 100 meters of a target. Even armed with a half-ton of high explosives, such a weapon has little probability of destroying a bridge, communications center, runway, or aircraft shelter.

Still, conventionally armed missiles are effective "terror weapons" when aimed at cities. From the German V-2 missile attacks on London in World War II to the missiles used in the Iran-Iraq war and the Gulf War, these weapons have terrorized civilians. Partially effective ground-based defenses against such attacks may become more feasible as the technology advances. But since this type of attack can't reach the United States, should U.S. taxpayers pay for protecting allied cities from it?

A tactical ballistic missile with a nuclear warhead could destroy military targets such as airfields or communications and transportation nodes, but the 1987 INF Treaty between the United States and the Soviet Union has banned such intermediate-range nuclear weapons from the arsenals of the two nations. And although new nuclear countries could develop these weapons, it is doubtful they would use them against U.S. forces, given the ability of the United States to respond devastatingly.

In any case, defenses against nuclear tactical ballistic missiles would be technically difficult. Ground-based defenses would face the same problem that defeated the ABM systems of the 1960s and 1970s: a nuclear detonation preceding the main attack could blind their radars. And GPALS' reach isn't low enough to knock out tactical ballistic missiles, which barely leave the atmosphere. SDIO director Henry Cooper has testified that the system's "brilliant pebbles"—orbiting self-contained missiles that would detect and attack ballistic missiles on their own—couldn't hit missiles at altitudes below 100 kilometers.

One can conclude that the United States needs no further defense against tactical ballistic missiles: conventional warheads aren't a significant military threat; nuclear warheads, if ever built, wouldn't be used against U.S. forces abroad for fear of retaliation. Thus to invoke the threat from such missiles as justification for pouring money into theater defenses is simply irrational.

The Accident Threat

That brings us to the last threat, and the one that merits the most careful consideration. Even after successful negotiations to reduce their strategic nuclear arsenals, both the United States and the Soviet Union will each retain between 6,000 and 9,000 intercontinental nuclear weapons through the end of the century and perhaps beyond; 4,900 of them on either side will be ballistic-missile warheads. Even if negotiations lead to further reductions, it is doubtful that either nation will reduce its arsenal to under 1,000 warheads in the foreseeable future.

Yet even as the number of nuclear weapons and the probability of their deliberate use declines, the risk of either an accidental or an unauthorized launch of strategic missiles remains. In an accident, a launch crew is convinced it has a valid order to act when in fact none has been issued. A missile launch is unauthorized if the crew acts without a command from the proper national authorities.

These two scenarios demand proportionately more attention as the prospects of a deliberate nuclear exchange diminish. Bomber-delivered nuclear weapons can be recalled or intercepted hours after their dispatch, but neither ballistic missiles nor long-range nuclear cruise missiles are recallable. Remote though it may be, the possibility of an unsanctioned launch of one or more nuclear delivery vehicles is real and presents the only danger of a nuclear attack on the United States.

The probability of a peacetime accidental launch of a U.S. or Soviet land-based missile is very small, but it can increase somewhat during practice exercises. And during a deepening crisis, when controls are relaxed and launch crews are under enormous stress, an accidental launch of one or more such weapons is not unthinkable. . . .

Measures to Avoid Catastrophe

There are two conceivable ways to avoid the catastrophe of accidental or unauthorized launches of nuclear ballistic missiles.

The first is an ABM system, based either on the ground or in space, that shoots missiles down as they approach the United States. In January 1988, Sen. Sam Nunn (D-Ga.), chair of the Armed Services Committee, suggested that the United States redirect SDI research toward developing a limited shield against a potential accidental or. unauthorized ballistic-missile launch. His proposed "accidental launch protection system" (ALPS) would use land-based—as opposed to Star Wars' space-based—high-acceleration interceptor missiles to destroy a few warheads as they reenter the atmosphere. At the end of July 1991, the Senate voted 60-39 to fund essentially this proposal. . . .

The Fading Third World Threat

GPALS [Global Protection Against Limited Strikes] is an extensive system of space and ground-based interceptors to defend the United States, its allies, and its troops abroad from third-world missile attacks and unauthorized or accidental launches from the Soviet Union. . . .

The threat GPALS is supposed to address may not be growing. The Gulf war demonstrated that countries like Iraq pose a missile threat. However, most third-world missiles are short-ranged, use outdated technology, and were bought abroad, not built at home. Missile suppliers will be scarce in the future, particularly if China continues to exercise the restraint it began after its transfer of missiles to Saudi Arabia in 1988. Advanced third-world attempts to "home-grow" these weapons, for example in Argentina and Brazil, are faltering.

Joel Wit, *The Christian Science Monitor*, April 2, 1991.

Not only would the ALPS deployment require renegotiating and probably drastically changing the ABM Treaty, but it could also arrest or reverse efforts to reduce the nuclear arsenals of the superpowers. The Soviet Union could perceive an ABM deployment in the United States as a threat to *its* nuclear deterrent and might then proceed to build more warheads to ensure that its weapons could penetrate U.S. defenses.

Nor could such a system assuredly shoot down every rogue warhead aimed at the United States. The system would be especially vulnerable to low-flying, submarine-launched ballistic missiles and cruise missiles launched close to U.S. shores.

For its part, SDIO has proposed assigning defenses against accidental or unauthorized nuclear attacks to the GPALS system of about a thousand of the so-called brilliant pebbles. However, such a system would offer little protection for two reasons. First, the brilliant pebbles would have to be turned off during peacetime to avoid attacking civilian space launches. Since an accidental or unauthorized launch can't be foreseen, it could occur while GPALS was on non-alert status. Second, only a few brilliant pebbles would be in range of an accidental or unauthorized launch at any time. At best, they could intercept only a fraction of the rogue weapons released by, say, a submarine. And, as with the ALPS proposal, deploying GPALS would destroy the ABM Treaty, an eventuality some Star Wars supporters might welcome but one most people would oppose. Finally, SDIO estimates the cost of GPALS to be about $40 billion, but estimates by independent experts put it at twice that price.

Thus neither ground-based nor space-based ballistic missile defenses can deal effectively with the threat of an accidental or unauthorized attack against the United States. Fortunately, a sound alternative exists.

A Fail-Safe Solution

The difficulties of erecting an effective antiballistic-missile system to counter this last type of threat have prompted the emergence of another solution: simple self-destruct devices installed on all ballistic missiles of all nations. Just as NASA safety officers can, and often do, destroy civilian space launches that go awry, so can ballistic missiles carrying nuclear weapons be destroyed remotely.

This approach has two requirements—one technical, one diplomatic. The first is that an opponent couldn't exploit the self-destruct system to blunt an intended nuclear attack. The second is that all nations with nuclear ballistic missiles adopt such self-destruct mechanisms. . . .

So far at least, no military or government officials have raised objections to a self-destruct system for ballistic missiles. The technology for this solution exists; diplomatic negotiation to adopt it is the rational next step.

"*Effective use of . . . [space] for military purposes . . . may be needed to safeguard national interests.*"

Space Warfare May Become Necessary

John M. Collins

In the following viewpoint, John M. Collins argues that future space warfare may become necessary to protect countries' vital interests. Collins asserts that as nations explore space for raw materials and energy, a superpower nation or an international alliance could attempt to monopolize such resources, thus triggering hostilities with one or more nations with similar interests. The author also states that as nations increase their military interests in space, warfare in space could occur. Collins is a defense specialist for the Congressional Research Service, the research branch of Congress, in Washington, D.C.

As you read, consider the following questions:

1. Why does Collins believe that outer space is important to military operations on earth?
2. In the author's opinion, how can nonviolent wartime tactics in space be effective?
3. How could a nation's military control of space during a war ensure victory, according to Collins?

Adapted from *Military Space Forces: The Next Fifty Years* by John M. Collins. Washington, DC: Congressional Research Service, 1989.

Every country and coalition on Earth has political, economic, military, social, and scientific interests in space. Survival, by definition, is the only vital interest. Physical security, prosperity, power, progress, and freedom of action are universally important. Additional interests of similar magnitude, such as peace and stability, appeal to some, while others spurn them or assign low priorities.

Nearly every interest in space has potential security implications. Economic competition on the moon or Mars, for instance, could cause war. Scientific probes to advance understanding of our universe could have serendipitous consequences comparable to Einstein's Special Theory of Relativity, which unexpectedly paved the way for nuclear weapons.

This study singles out three competitive, perhaps incompatible, interests with sweeping significance: political cooperation; economic exploitation; military power.

Political Cooperation

Political cooperation, the most widely professed interest in space, conforms with the Charter of the United Nations (UN), which seeks to create "conditions of stability and well-being" that must accompany "peaceful and friendly" international relations. Ninety-six nations subscribe to the Outer Space Treaty of 1967. Article I specifies that exploration and other endeavors "shall be carried out for the benefit . . . of all mankind." Article II disapproves sovereignty anywhere in space. The Moon Treaty of 1979 has attracted only seven parties, none of whom now have space capabilities, but its explicit restrictions seem to reflect prevailing UN views: "Neither the surface nor the subsurface of the moon [or other celestial bodies within the solar system] . . . shall become property" of any person, state, or other organization.

National leaders generally live in a less altruistic world, but openly proclaim that collaboration serves the common good better than zero-sum games with only one winner. They also agree that community efforts are preferable to hegemony by any power in space, if public statements reflect true sentiments. Responsible authorities rarely release opinions that oppose equal opportunities to share rewards and risks. The strength of such convictions will be tested when economic competition quickens in space.

Economic Exploitation

Economic interests in space center on vast and diversified raw materials, inexpensive solar power, and industrial techniques made possible by near vacuum and low gravity. Prospects for extraordinary productivity and growth are within relatively easy reach of Planet Earth.

The lunar mantle contains all essential elements for the full range of manufacturing and construction on a grand scale, plus so much oxygen that some call the moon "a tank farm in space." Smelters could rely less on costly chemical treatments, because heat alone would remove most impurities from many ores in that environment. Waterless cement, specialty alloys, exotic composites (such as glass/metal mixtures stronger than steel, yet transparent as crystal), powder metallurgy, cold welding, free fall casting, and superconductors represent a few among many processes and products that plants in space could facilitate. "Sanitation engineers" could dispatch unconvertible waste on trajectories that collide with the sun or disappear in deep space.

Interests in economic exploitation of space are intrinsically neither virtuous nor evil. Efforts that benefit the world community could foster strategic stability. Peaceful competition, such as that in Antarctica (where seven countries have staked out territorial claims), is theoretically possible. Efforts by any superpower or collection of states to monopolize the bounties of space and couple them with earthbound assets, however, might trigger international strife, because success in some circumstances could sensationally upset the global balance of power.

Military Power

Space, the ultimate "high ground," overarches Planet Earth, its occupants, and all activities thereon. Effective use of that medium for military purposes therefore may be needed to safeguard national interests in survival, security, peace, power, stability, and freedom of action.

Every technologically advanced land, sea, and air service already depends on space satellites to such a degree that traditional command, control, communications, and intelligence (C3I) skills may languish, much like pocket calculators made slide rule proficiency rare. Reliance continues to increase, because systems in space offer strategic and tactical advantages that are otherwise unavailable: national technical means of verification for treaty compliance and crisis monitoring; geodetic surveys to assist military map makers and target planners; weather prediction; early warning and post-attack assessment; nuclear detonation detection; global positioning/navigation data; and observable order of battle information. Satellites relay most military intercontinental telecommunications and an increasing share of tactical traffic.

Military interests in space almost surely will intensify and spread during the next decade. How smoothly they will mesh with aforementioned interests in political cooperation and economic exploitation is problematic. Reconciliation and collision

159

are divergent possibilities. Plans that address both extremes and contingencies between seem advisable, pending clarification. . . .

Projected Capabilities

Superiority in space could culminate in bloodless total victory, if lagging powers could neither cope nor catch up technologically. Lesser unilateral breakthroughs could reduce rival abilities to resist aggression on Earth and beyond in many significant ways.

Space Is Crucial to Military Success

As long as there is conflict between competing interests and ideologies, space will not be exempt from it. Indeed, we are entering an era when space control is becoming *the* crucial military leverage, and may determine the course of future conflicts—without a shot ever being fired by terrestrial forces.

Military forces have historically opened the way into new frontiers of human endeavor, whether it was navies opening up the high seas, or, as in our own history, the Army exploring the new Western frontier, and providing security for settlers, homesteaders and railroads.

There is no reason the development of the space frontier should not follow the same pattern.

Malcolm Wallop, *The Wall Street Journal*, August 2, 1990.

Science and technology are twin keys to future space capabilities, but forecasters in those fields find it troublesome to predict the progress of friends, much less that of opponents whose exploratory programs are well concealed. Pundits who insist that any technological problem is insolvable have repeatedly been proven wrong. Cracked crystal balls invariably overlook impending developments of great magnitude. *Technological Trends and National Policy*, a 1937 U.S. study, failed to foresee radar, jet engines, and nuclear weapons, which were operational within eight years or less. Dr. Vannevar Bush, Director of Scientific Research & Development, and the Von Karmann report entitled *New Horizons* both discounted ICBMs in 1945; Soviet tests took place in 1957. Skeptics in 1961 doubted men would soon land on the moon and return safely during that decade. They were mistaken.

Since surefire predictions perhaps are impossible, given the dearth of hard data, it is important to press states of art wherever technological surprise in or from space conceivably could alter the military balance decisively. Failure to do so would deprive decisionmakers of sufficient vision to determine what enemy capabilities are possible in any given time frame, separate possibilities from probabilities, and take appropriate action. . . .

Hard Kill Weapons

Planners must pick the most appropriate arms from a diversified arsenal. Mass destruction and precision instruments compete for attention. Directed energy weapons (DEW) and some types of radiation attack at the speed of light. Space mines move sluggishly in comparison. Spoofing normally is a nonlethal option, but might misdirect space weapons against friendly forces. Target accessibility and survivability, concerns for escalation, and damage desired strongly influence decisions to engage any point or area target with particular weapons.

"Hard kill" weapons forcibly break the surface of targets, then damage or destroy their contents. Violence is evident to observers. Explosives and kinetic projectiles are representative instruments.

Nuclear warheads are the most escalatory of all area weapons. Space-to-surface and surface-to-surface shots that detonate on Earth or in its atmosphere would be equally lethal, but space-to-surface delivery currently is (may always be) more costly, less accurate, and tougher to control. Nuclear hard kill capabilities would seldom be cost-effective against targets in space, because vacuum restricts blast and heat radii so severely that conventional explosives and kinetic energy weapons can accomplish most missions equally well at less expense. Soft kill nuclear radiation, unimpeded by atmosphere, conversely could cover orders of magnitude more volume than near Earth's nap. It would work especially well against suspected targets in low earth orbit (LEO) but less consistently in deep space, unless targets were located more accurately, since the lethal range of radiation is limited, even in a vacuum. Nuclear radiation in any event suffers one great disadvantage: it cannot distinguish friend from foe. Electromagnetic pulse, for example, might "wound" users as grievously or worse than intended victims.

High-Speed Energy Weapons

Conventional explosives and kinetic energy weapons (KEWs) are designed exclusively for hard kills. Proper employment in space depends extensively on precise targeting intelligence, plus "smart" (even "brilliant") munitions. Direct hits are almost obligatory for conventional explosives, because vacuum cancels the concussive effects that make many near misses deadly on Earth. The tolerable margin of delivery error also is small for KEWs, even "shotgun" styles that scatter pellets in the orbital paths of speeding targets. Offensive KEWs plummeting from space-to-

Earth at Mach 12 or more with terrific penetration power have a marked advantage over defensive Earth-to-space counterparts that accelerate slowly while they fight to overcome gravity. Simple weapons associated with sabotage and other special operations open low key hard kill offensive options for use against space installations anywhere in the Earth-Moon System.

Military Protection in Space

Space will affect all aspects of our lives in the next century. Just as on the high seas, space-related peaceful activities will only flourish if assured of the backing of an on-call military capability to protect our space assets. An American capability to operate beyond Earth's atmosphere without interference will help ensure peaceful use of space by all friendly nations.

Edward L. Rowny, *Los Angeles Times*, July 18, 1989.

Speed-of-light directed energy weapons (DEW) likely will be preferred implements whenever space-to-space attackers try to take point targets by surprise. Weapon quality lasers and particle beams await substantial technological breakthroughs, however, before they can operate successfully across the boundary between atmosphere and space. Space-*toward*-Earth capabilities against targets in rarified atmosphere are simplest to perfect. Space-*to*-Earth DEW operations are much farther in the future, because dense air is more difficult to penetrate and targets ashore or afloat on the surface are easier to shield. R&D specialists anticipate Earth-to-space DEWs at an earlier date, since large power supplies are more readily attainable on terra firma than on spacecraft and most targets in space will always be relatively soft. One undesirable attribute merits mention: the source of super fast laser attacks that bounce off space-based reflecters may be untraceable. Catalytic conflicts and retaliation against wrong parties could result. Space strategists concerned with unplanned escalation and other unwanted contingencies should consider such possibilities, and employ DEWs cautiously.

Soft Kill Weapons

"Soft kill" weapons penetrate target surfaces without impairing them, then selectively disorient, damage, or destroy humans and/or sensitive mechanisms inside. Electronic countermeasures and nuclear radiation are representative instruments.

Soft kills generally cost less and escalate conflict less than hard kills. Jamming, which reduces communication data rates or renders signals unintelligible, requires complex techniques and sophisticated equipment. Other potential methods are simple. Spray paint on satellite camera lenses, blinding light on laser reflectors, and surreptitious introduction of foreign objects into booster fuel are typical possibilities. False commands and other forms of "spoofing" may cause enemy satellites to malfunction. Victims who suspect foul play are hard pressed to prove it, if opponents are clever.

Armed forces in space, as on Earth, must concentrate physical presence or firepower to defeat enemies. Simultaneous assaults to decrease warning times and/or increase shock effects on widely separated targets, however, are tricky propositions, particularly when weapons have dissimilar characteristics and firing points are far apart. Directed energy weapons promise to simplify, but cannot solve, such problems. Opportune maneuvers, coupled with surprise and deception, always will be important. . . .

Former Secretary of State Henry A. Kissinger, at a March 1974 press conference in Moscow, asked "What in God's name is strategic superiority?" It may be unilateral control of space, which overarches Planet Earth, all occupants, and its entire contents. If so, possessors of that vantage position could overpower every opponent. They might, in fact, impose their will without fighting, a feat that Sun Tzu called "the acme of skill" 25 centuries ago. U.S. military space forces therefore need means to forestall strategic surprise from that quarter and respond successfully, unless best case estimates prove correct as events unfold.

"The elimination of space weapons is of paramount importance."

Space Should Be Used for Peaceful Purposes

Sriharaburti Chandrashekar

Many scientists and religious leaders urge that space be used solely for peaceful purposes. Sriharaburti Chandrashekar, the author of the following viewpoint, agrees. Chandrashekar argues that advancing technology and the rise of space weapon systems, such as the SDI program, threaten peace in outer space. Chandrashekar believes that the lack of international rules governing the use of space will result in its further militarization. To keep space a peaceful region, the author asserts that a comprehensive ban of such weapons is needed. Chandrashekar is a researcher for the Launch Vehicle Program Organization in Bangalore, India.

As you read, consider the following questions:

1. Why does Chandrashekar believe it is important to impose a standard minimum altitude as a definition of space?
2. According to the author, why do ballistic missiles pose difficulties in establishing a definition for "space weapon"?
3. How can a space weapons ban be fair to all countries, according to Chandrashekar?

Excerpted from "Problems of Definition: A View of an Emerging Space Power," by Sriharaburti Chandrashekar, from *Peaceful and Non-Peaceful Uses of Space*, edited by the United Nations Institute for Disarmament Research, Geneva, Switzerland. New York: Taylor & Francis, Inc. 1991. Reprinted with permission.

The absence of a clear international regime governing the "peaceful" and "non-peaceful" uses of outer space poses major contradictions between domestic and international policies to an emerging space power. Coupled with the lack of progress in achieving genuine "peaceful" uses of space, this will eventually lead to more countries using space for military purposes. The word "peaceful" has two essential components associated with it. These are an absence of force or conflict and the presence of calm and tranquility. The current international regime of space is not a peaceful regime. Activities related to defensive weapons and support military functions can be carried out in outer space.

A categorization of the different uses of space along with the legal regimes associated with them indicated that activities related to the weaponization of space are a major area of concern requiring immediate initiatives. . . .

Banning Space Weapons

Different approaches are possible toward the elimination of space weapons. In one approach a generic ban could be imposed. This would logically require a definition of space. In another approach, a ban on ASAT [antisatellite] and BMD [ballistic missile defense] weapons only could be imposed. This would involve definition of "an object in space." To categorize objects in space, the orbit characteristic of a satellite, and a "cut-off velocity," and "time of flight" criteria for a ballistic missile could be used. Other variants of these basic approaches could also be considered.

An international verification mechanism that could include tracking of space objects, monitoring telemetry, and independent observation satellites, could monitor non-peaceful activity. Parameters such as radiation hardening, weight, power, nature of telemetry transmission, free availability of data and satellite services, and international participation could be used as additional elements to categorize peaceful uses.

The word "peaceful" can no doubt be defined to facilitate progress in the peaceful uses of outer space, but the approach involves the clear and unambiguous definition of the term "non-peaceful."

The major elements of such a definition should include:

- The use of force or the threat of use of force by or against a space object.
- The use of a space weapon.
- The use of space objects to aid and assist in military operations.

History of Space Weapons

The question of peace in space and the peaceful uses of outer space has been a major issue before the international community. Several excellent reviews have been published.

"As you can see, all this is for the sake
of your safety!"

N. Shcherbakov/*New Times*. Reprinted with permission.

So long as there were only two players in the space race,
geopolitical interests overshadowed other considerations and
military uses emerged as a major requirement. Early warning,
reconnaissance, navigation, and military communications were
the early drivers of the military effort. The first nuclear tests in
space took place in 1962. It is reasonable to conclude that the
resulting damage to several satellites was at least partly responsible for the prohibition of nuclear explosions in outer space as
enunciated in the Partial Test Ban Treaty and reemphasized later
in the Outer Space Treaty (OST) of 1967.

At the United Nations consensus was finally achieved on the
contents and text of a treaty on the activities of states in outer

space, which came into force in 1967.

The late 1960s and the early 1970s witnessed new developments, including new weapon systems for use in and against objects in space and on earth.

SDI and the Arms Race

The question of arms in space and the peaceful and non-peaceful uses of outer space became particularly important after the initiation of SDI in 1983. Since then space militarization, space weaponization, peaceful and non-peaceful uses of outer space, and the extension of the arms race into space have all become terms much used, discussed, and elaborated upon in several international fora.

However, in spite of the rhetoric and the plethora of viewpoints on approaches, little progress has been achieved in realizing the ideal goal of "the exploration and use of outer space for peaceful purposes," which is enshrined in the preamble to OST.

Recent progress in arms limitations talks between the superpowers has re-kindled hopes that progress on the ticklish issues of peaceful/non-peaceful uses of outer space is indeed possible. . . .

Importance of Banning Weapons

To an emerging space power, the issue of the elimination of space weapons is of paramount importance. There are, of course, several routes through which this goal can be realized. Each route has associated with it problems of definition of a number of critical terms.

The first approach is one through which all categories of "space weapons" can be eliminated. A typical example of this generic approach is the Antarctic Treaty. It prohibits all military activities in Antarctica. The geographical area covered is also clearly defined in the treaty as the area lying south of 60°S latitude.

A similar generic approach prohibiting the development, testing and stationing of space weapons would be a major step toward realizing a peaceful use of space.

To come to terms with this approach a clear unambiguous definition of "a space weapon" is needed. Such a definition should be comprehensive enough to cover both ASAT weapons as well as BMD weapons, which are the major sources of concern to the international community today.

Defining Space Weapons

The characteristics of weapons in general and space weapons in particular are:
- Like all weapons, space weapons should be capable of destroying, damaging, or otherwise interfering with the normal functioning of the object that is the target of attack.

- Unlike other conventional weapons, a space weapon (which could also be based on land, sea, or in air) is directed against an object in space. This object could be a satellite, or a ballistic missile flying through space or its nuclear warhead.
- Unlike other weapons, a space weapon could be stationed in space or flying a trajectory through space and directed against objects on the earth, in the air, or in space.

Where Outer Space Begins

Because one of the characteristics of a space weapon is that it is directed against an "object in space," it is necessary to enunciate rather clearly when an "object is in space." One method of doing this, which is rather obvious, is to delineate or define space. To draw a parallel with the Antarctic Treaty, we need to define the equivalent of 60°S latitude for "space."

Many approaches have been thought of to delineate space. One approach to this definition imposes a minimum altitude limit (a 100-km altitude is often suggested as the lowest boundary of space). This approach involving a clear spatial delineation has much merit because all foreseeable ASAT and BMD weapons (at least the ones not covered by the existing treaties) would be clearly covered by this definition. Any object that is therefore at an altitude of more than 100 km would qualify as an object in space. Satellites, relevant portions of the boost, post-boost, and midcourse trajectories of ballistic missiles would be covered. To cover pop-up systems, "weapons stationed in space" could be defined to include weapons whose trajectories are in practice above the specified altitude. This altitude could be fixed with respect to any of the standard earth models that are available such as the Goddard earth models.

The advantage with this approach is that it is clear, simple, and unambiguous. Verification problems are to some extent simplified.

Problems of Defining Altitude

However, in spite of extensive debate in the UNCOPUOS [United Nations Committee on the Peaceful Uses of Outer Space], the spatial approach toward the definition of space has eluded consensus. Other definitions including functional ones have also eluded consensus. The major reason is not because the definition of space is difficult. The problem arises largely because of the strategic concerns of the space powers, especially, since ballistic missiles fly a large part of their trajectory through space. These concerns are understandable in UNCOPUOS, which is not supposed to deal with disarmament questions, but a spatial definition of space may not pose any special difficulties to the space powers in a multilateral disarmament forum. This is so, because the subject of discussions are space weapons and the issues addressed are strategic issues. In addition, the status of ballistic missiles flying through space is one of the major items of interest for arms control discussions. Therefore an altitude limit definition for the lower boundary of space, such as a 100 km lower limit, should not be ruled out just because there has been no agreement in UNCOPUOS.

A Peaceful Region for Everyone

Existing bodies of law suggest ways to protect the "high seas" of space and the celestial islands of the cosmos. The 1967 Outer Space Treaty, for instance, includes a number of "wilderness provisions" including freedom of scientific investigation, public access, a ban on "nuclear weapons or any other weapons of mass destruction," noncontamination of celestial environments and the internationalization of space as "the province of all mankind."

Gar Smith, *Earth Island Journal*, Winter 1987.

In case the altitude limit definition of space runs into problems, are any alternative approaches possible? One possible approach would be not to define space directly but rather try to define space indirectly by defining "objects in space"—the subject matter of space weapons. Since objects in space are both targets as well as weapon systems, the definition of an "object in space" should be comprehensive enough to include the weapon systems of immediate interest—ASAT and BMD weapons. This poses additional definition problems, which, though not difficult to overcome by themselves, make the issue of definition more complex.

ASAT systems are directed against satellites, so "the object in space" is a satellite that is in orbit around the earth. This characteristic, namely "in orbit around the earth," can be used to define one characteristic of an object in space. This would take care of weapon systems directed against satellites and weapon systems in orbit around the earth.

The "orbit characteristic" definition of an "object in space" does not take into account objects that spend only a portion of their trajectories in space. Because this is a characteristic of both ballistic missiles and some weapon systems such as pop-up systems, ways of overcoming this lacuna have to be found. . . .

Distinguishing Weapon Systems

The third approach toward the question of weapons in space would be to consider each kind of weapon system separately. In this approach ASAT weapons would be treated separately and BMD weapons separately. Other categories of space weapons would be dealt with as and when they pose a threat. Definition problems in this mode would be less difficult. The targets of interest could be easily defined for an ASAT treaty. Similarly, a ballistic missile can be defined and weapons directed against them banned. This approach of separating ASAT and BMD weapons would, however, pose some special problems. This arises from the close technology coupling between ASAT and BMD weapons. This linkage makes it difficult to verify a selective ban on only one of either ASAT or BMD weapons. This would not be a preferred approach to take unless the international community wants to limit only one kind of weapon system or in some way the ASAT and BMD weapon systems are treated as a package. If the package approach is adopted, it would practically amount to an approach similar to the second approach suggested earlier.

The fourth approach would be to take each weapon type separately and deal with it. Thus kinetic energy weapons, beam weapons, particle beam weapons etc., which serve both ASAT and BMD functions, could be dealt with separately. These approaches would try to use technical characteristics such as beam energies, laser pulse duration, fluence, kinetic energies or velocities, to achieve the required objectives of defining, when their use could be considered a weapon. The targets themselves could also be identified as a satellite or a ballistic missile using functional elements outlined whether in approach one or approach two.

In all approaches it is mandatory that components of weapon systems are also defined.

Ideal Solutions

To an emerging space power the first approach where a clear spatial definition of space is available would be preferred. This approach is generic, and it has the advantage that it completely eliminates the possibility of development of space weapons by all countries. A clear delineation of space also facilitates verification.

The second approach, although not so comprehensive (only ASAT and BMD weapons are covered), would still be acceptable if problems associated with definitions could be resolved. A number of approaches toward evolving these definitions exist. However, this is not as comprehensive an approach as approach one. Since ASAT and BMD weapons are the weapons of immediate concern, this could be a pragmatic approach to take, if the definition approaches suggested either singly or in combination can be used to characterize an object in space.

The third approach of separating ASAT and BMD weapons should be avoided at all costs in view of the close technology coupling the two classes of weapons. Eliminating one without

170

eliminating the other would cause more problems and pose difficult problems of verification.

The fourth approach of treating the weapons by type (such as kinetic energy weapons, beam weapons, particle weapons) would also not be acceptable, because by selective banning of some categories, international security is not enhanced. Emerging space powers may also not adhere to such a treaty because they would deem it discriminatory. In the long term it would create precisely the same concerns that have happened in the area of nuclear weapons and missiles—proliferation concerns, prohibitions on technology, and discriminatory embargo regimes.

Peace Requires a Total Ban

To an emerging space power, space is still a weapon-free arena. A comprehensive ban on weapons would provide a truly non-discriminatory regime for all countries. Selective banning of space weapons or regimes that regulate the number of weapons etc. would result in a discriminatory regime. Whereas such a regime may be successful in the short term, national compulsions could force many emerging space powers into embarking on their own space weapons program. Problems similar to what is happening in the area of nuclear weapons, missiles etc., would repeat themselves. Whereas individual short-term self-interests of space powers would be preserved by a selective approach toward elimination of space weapons, the long-term interests of the whole world would be jeopardized since the road to embark on a space weapons program is equally open to all countries. The resultant loser is, of course, world peace and world security.

Understanding Words in Context

Readers occasionally come across words they do not recognize. And frequently, because they do not know a word or words, they will not fully understand the passage being read. Obviously, the reader can look up an unfamiliar word in a dictionary. By carefully examining the word in the context in which it is used, however, the word's meaning can often be determined. A careful reader may find clues to the meaning of the word in surrounding words, ideas, and attitudes.

Below are excerpts from the viewpoints in this chapter. In each excerpt, one of the words is printed in italics. Try to determine the meaning of each word by reading the excerpt. Under each excerpt you will find four definitions for the italicized word. Choose the one that is closest to your understanding of the word.

Finally, use a dictionary to see how well you have understood the words in context. It will be helpful to discuss with others the clues that helped you decide on each word's meaning.

1. The worldwide condemnation of Iraq's devastation of Kuwait reinforces the status of Iraq as an international *PARIAH.*

   PARIAH means:

   a) leader   c) peacemaker
   b) outcast   d) debtor

2. Peace and stability in outer space appeal to some nations, but others *SPURN* these interests.

   SPURN means:

   a) reject   c) applaud
   b) complicate d) hide

3. The increase in current Third World missile capabilities is a *HARBINGER* of the complexities America's military will face in the next century.

   HARBINGER means:

   a) guideline
   b) procedure
   c) simulation
   d) herald

4. The United Nations argues that exploration of the Moon should be conducted for the benefit of all humanity. However, leaders of powerful nations are less *ALTRUISTIC*.

   ALTRUISTIC means:

   a) worthy
   b) unselfish
   c) weak
   d) simplistic

5. The idea of using space weapons to protect the entire United States from a nuclear attack has been ridiculed by many scientists. They believe total protection is an *INFEASIBLE* proposition.

   INFEASIBLE means:

   a) untimely
   b) technical
   c) worthwhile
   d) unworkable

6. Domination of earth's orbit, the Moon, and other regions of space by one nation would be a *UNILATERAL* control of space.

   UNILATERAL means:

   a) one-sided
   b) powerful
   c) justifiable
   d) reckless

7. The growing number of Third World nations that conduct missile tests have made Americans aware of the *PROLIFERATION* of ballistic missiles.

   PROLIFERATION means:

   a) multiplication
   b) benefit
   c) limitation
   d) secret

8. While Iraq did not use chemical weapons in the Persian Gulf War, the *SPECTRE* of their use was a constant source of unease to Allied forces.

   SPECTRE means:

   a) nagging doubt
   b) false alarm
   c) haunting image
   d) secret intent

Periodical Bibliography

The following articles have been selected to supplement the diverse views presented in this chapter.

Stephen Aubin	"The Patriot and SDI: Changing Images," *The World & I*, May 1991.
Tom Bethell	"Star-Wars Wars," *The American Spectator*, April 1991.
Richard H. Buenneke Jr.	"After the Storm: Preparing for the Next Space War." *Final Frontier*, March/April 1992. Available from PO Box 534, Mt. Morris, IL 61054-7852.
Jeffrey Denny	"Star Struck," *Common Cause Magazine*, March/April 1991.
Daniel O. Graham	"An SDI Program for the Clear and Present Danger," *New Dimensions*, May 1991. Available from PO Box 811, Grants Pass, OR 97526.
Miles Harvey	"Star Wars: The Sequel," *Utne Reader*, January/February 1992.
Jonathan Jacky	"Throwing Stones at 'Brilliant Pebbles,'" *Technology Review*, October 1989.
Edgar D. Mitchell	"An Astronaut Against Star Wars," *The Progressive*, December 1990.
Meera Nanda	"A New Star Wars for Bush's New World Order," *Guardian*, February 13, 1991.
John L. Piotrowski	"SDI and Missile Proliferation," *Global Affairs*, Spring 1991. Available from International Security Council, 1155 15th St. NW, Washington, DC 20005.
Aldric Saucier	"Lost in Space," *The New York Times*, March 9, 1992.
Phyllis Schlafly	"Soviet Breakup Proves We Need SDI," *Conservative Chronicle*, October 9, 1991. Available from PO Box 11297, Des Moines, IA 50340-1297.
Baker Spring	"Mad Dogs and Congressmen: Arguments Against SDI Are Impotent and Obsolete," *Policy Review*, Fall 1991.
Edward Teller	"From Brilliant Pebbles to Brilliant Eyes," *The New York Times*, June 23, 1991.
Malcolm Wallop	"Congress Forgets the Meaning of Defense," *The Wall Street Journal*, August 2, 1990.

Glossary

ABM   antiballistic missile; a missile used to intercept and destroy a guided missile.

ASAT   antisatellite weapon.

biosphere   the part of the earth's crust, waters, and atmosphere where living organisms can survive.

BMD   ballistic missile defense.

booster   a rocket used to give extra power during liftoff or at another stage of a spacecraft's flight.

chlorofluorocarbons (CFCs)   substances composed of atoms of chlorine, fluorine, and carbon; CFCs, which are used commercially as coolants and in other products, damage earth's ozone layer, which shields the earth from ultraviolet radiation, and also contribute to the greenhouse effect.

DOD   Department of Defense.

DOE   Department of Energy.

DOT   Department of Transportation.

EOS   Earth Observing System.

ESA   European Space Agency, a thirteen-nation group.

exobiology   the study of life on other planets.

galaxy   a large system of stars held together by mutual gravitation.

GEO   geosynchronous orbit; see *geosynchronous.*

GPALS   Global Protection Against Limited Strikes; a proposed program of ground-based and space-based weapons defense systems.

geosynchronous   the orbit of a spacecraft or **satellite** traveling at the same speed as the earth's rotation, so that the orbiting object remains above the same point on the earth; for example, to regularly receive and transmit signals, communications satellites must remain in geosynchronous orbit.

HE-3   an isotope of **helium** found in lunar soil.

helium   the second lightest chemical element and the second most abundant; produced in stars by the fusion of hydrogen atoms.

HLLV   heavy-lift launch vehicle; a proposed alternative to **STS.**

HST   Hubble Space Telescope.

hypersonic   refers to airflight five or more times greater than the speed of sound.

ICBM   intercontinental ballistic missile.

laser   a device that produces a narrow beam of high-intensity radiation at different wavelengths; this beam can theoretically be used as a space weapon.

launcher   a rocket vehicle used to put a satellite or spacecraft into orbit.

LEO   low earth orbit.

light-year   the distance traveled by light in one year, about 5,880,000,000,000 miles.

LOX liquid oxygen.

Malthusian theory theory of English economist Thomas R. Malthus which states that population tends to increase at a faster rate than its means of subsistence and that unless it is checked by moral restraint or by disease, war, or other disaster, widespread poverty and degradation inevitably result.

microgravity an extremely low level of gravity; for example, the gravity experienced by shuttle crews is one-millionth the gravity on earth's surface.

module a single section of a spacecraft that can be separated from other sections.

NASA National Aeronautics and Space Administration.

NASP National Aerospace Plane, proposed alternative to the space shuttle.

NOAA National Oceanic and Atmospheric Administration.

orbit the path of an object, such as a moon or satellite, that revolves around another object, such as a planet.

OST Outer Space Treaty.

probe an automated, crewless spacecraft used to gather information or perform experiments in space or on other planets; it transmits its findings back to earth by radio waves.

propellant a chemical or chemical mixture burned to create thrust for a rocket or other spacecraft.

R&D research and development.

robotics the use of automated machines to replace human effort.

satellite any natural or artificial body in **orbit** around a planet; commonly used to refer to manufactured machines that orbit earth and send back information.

SDI Strategic Defense Initiative, also known as Star Wars; now being researched by the Pentagon, SDI is a defense plan that includes the use of space weapons.

SEI Space Exploration Initiative; proposed in 1989, SEI is a plan to send manned missions to the Moon and Mars in the twenty-first century.

SETI Search for Extraterrestrial Intelligence

single-stage to orbit reusable spacecraft which would take off either vertically or horizontally without added boosters.

SSF space station *Freedom.*

SSTO single-stage to orbit.

STS Space Transportation System, also known as the space shuttle.

terraforming the theoretical process of altering a planet's or moon's natural state to create conditions more like those on earth, making the planet suitable for human habitation.

thrust the push given a rocket by its engine.

Chronology of Events

1900	H.G. Wells publishes *The First Men in the Moon*.
1901	Radio pioneer Nikola Tesla concludes that strange, possibly artificial, signals from one of his radio receivers originated from Mars or Venus.
1919	Robert H. Goddard publishes *A Method of Reaching Extreme Altitudes*.
1923	Hermann Oberth publishes *The Rocket into Planetary Space*.
1926	Goddard fires first liquid propellant rocket.
1929	Oberth publishes *Roads to Space Travel*.
1930	American Interplanetary Society, predecessor of the American Institute of Aeronautics and Astronautics, is formed.
1931	Germany develops a 110-pound thrust rocket.
1934	Wernher von Braun, a German rocket scientist, tests a 660-pound thrust rocket.
1935	Goddard's gyrostabilized rocket reaches an altitude of 7,500 feet.
1942	Germany successfully test-fires its first V-2 rocket to a distance of fifty-three miles.
1945	United States Army fires its WAC Corporal rocket to a distance of 43.5 miles.
1949	United States launches a Bumper-WAC two-staged rocket 244 miles above earth.
1955	United States announces its Vanguard satellite program.
October 4, 1957	Soviet Union launches world's first satellite, Sputnik 1.
January 31, 1958	United States launches its first satellite, Explorer 1.
October 1, 1958	The National Aeronautics and Space Administration (NASA) is created.
October 11, 1958	NASA launches its first unmanned spacecraft, *Pioneer 1*, but an attempted lunar landing fails.
April 8, 1960	American astronomer Frank Drake begins Project Ozma, the world's first SETI program.

April 12, 1961	Soviet cosmonaut Yury Gagarin becomes the first human to orbit earth.
May 5, 1961	Alan Shepard, in a fifteen-minute suborbital flight aboard *Mercury Freedom 7*, becomes the first American to reach outer space.
May 25, 1961	President John F. Kennedy calls upon the United States to achieve a lunar landing by the end of the decade.
February 20, 1962	John Glenn becomes the first American to orbit the earth, circling it three times in *Friendship 7*.
June 16, 1963	Soviet cosmonaut Valentina Tereshkova becomes the first woman in space.
February 3, 1966	Soviet Union's unmanned *Luna 9* becomes the world's first spacecraft to land on the Moon.
June 2, 1966	United States achieves its first unmanned lunar landing with *Surveyor 1*.
January 27, 1967	A flash fire kills U.S. astronauts Virgil I. Grissom, Edward H. White, and Roger B. Chaffee at Cape Canaveral, Florida, during a countdown simulation of an Apollo mission.
July 20, 1969	*Apollo 11* crew members Neil Armstrong and Edwin Aldrin become the first astronauts to land on the Moon.
February 11, 1970	Japan launches its first satellite.
April 19, 1971	The world's first space station, *Salyut 1*, is launched by the Soviet Union.
May 14, 1973	United States launches *Skylab*, its first orbiting space station.
July 17, 1975	*Apollo 18* astronauts and Soviet cosmonauts conduct first joint U.S.-Soviet spacecraft linkup.
August 20, 1975	United States launches *Viking 1*, an unmanned Mars lander.
August 20 and September 5, 1977	United States launches *Voyager 2* and *Voyager 1* unmanned space probes.
July 11, 1979	*Skylab 1* falls out of earth's orbit and breaks up over Australia.
December 23, 1979	The European Space Agency (ESA) launches its first satellite aboard *Ariane 1*.
November 12, 1980	*Voyager 1* discovers new rings and moons of Saturn.
April 12, 1981	United States launches *Columbia*, the world's first space shuttle.

August 25, 1981	*Voyager 2* transmits highly detailed pictures of Saturn's rings and satellites. It continues on toward Uranus (1986) and Neptune (1989) and is expected to explore beyond the solar system along with *Voyager 1*.
March 23, 1983	U.S. president Ronald Reagan announces plans for the Strategic Defense Initiative (SDI), a space-based defensive measure against ballistic missiles.
November 28, 1983	Space shuttle *Columbia* carries the European Space Agency's *Spacelab* into orbit.
January 25, 1984	Reagan calls for the development of a permanently occupied space station within a decade.
June 10, 1984	U.S. Department of Defense destroys a mock ballistic missile warhead on its flight through space, the first such test with a ground-based interceptor.
January 28, 1986	Space shuttle *Challenger* explodes seventy-three seconds after liftoff, killing all seven crew members, including Christa McAuliffe, who was to be the first teacher in space.
February 20, 1986	Soviet Union launches the *Mir* space station.
October 10, 1986	SETI scientist Paul Horowitz of Harvard University uses a radio telescope and the equivalent of a supercomputer to lock onto what he believes is an extraterrestrial signal. After reaiming the telescope, the signal vanishes.
November 15, 1988	Soviet Union launches *Buran*, its first unmanned shuttle, which makes two orbits of the earth.
April 20, 1989	U.S. president George Bush establishes the National Space Council, a policy-making group headed by Vice President Dan Quayle.
July 20, 1989	On the twentieth anniversary of the first lunar landing, Bush announces the goals of returning humans to the Moon and sending them to Mars in the twenty-first century.
April 24, 1990	NASA launches the Hubble Space Telescope. Although its primary mirror was discovered to be flawed, NASA plans to correct the problem.
October 6, 1990	*Ulysses*, a joint U.S.-European spacecraft, is launched from space shuttle *Discovery* to probe unexplored regions above and below planets' orbits and the polar regions of the sun.

179

July 16, 1991	ESA launches ERS-1 (European Remote Sensing satellite) to monitor Earth's environment.
December 7, 1991	Ten of the eleven republics of the Commonwealth of Independent States (formerly the Soviet Union) agree to the creation of the Interstate Space Council, a new space agency.
October 12, 1992	NASA schedules this date, the 500th anniversary of Columbus's first landing in the Americas, to begin searching millions of radio frequencies for signals produced in other galaxies by intelligent life.
November 1995	NASA and the ESA plan to launch *Cassini*, the first probe to study systems of the planet Saturn in detail.
2002	ESA plans to launch *Hermes*, an unmanned space shuttle.

Organizations to Contact

The editors have compiled the following list of organizations that are concerned with the issues debated in this book. All have publications or information available for interested readers. For best results, allow as much time as possible for the organizations to respond. The descriptions below are derived from materials provided by the organizations. This list was compiled upon the date of publication. Names, addresses, and phone numbers of organizations are subject to change.

American Institute of Aeronautics and Astronautics (AIAA)
370 L'Enfant Promenade SW
Washington, DC 20024
(202) 646-7400

The AIAA is comprised of aeronautical and astronautical scientists and engineers, many of whom support manned space exploration, construction of space station *Freedom*, and a mission to Mars. AIAA publishes *Aerospace America*, a monthly magazine covering aeronautics and space technology.

Association of Space Explorers (ASE)
35 White St.
San Francisco, CA 94109
(415) 931-0585

ASE, a professional organization of individuals who have flown in space, encourages international cooperation in manned space exploration. It supports manned missions to the Moon and Mars by the United States and other spacefaring nations. ASE publishes *The SPACExplorer* newsletter quarterly, in addition to position papers and reports.

Center for Defense Information (CDI)
1500 Massachusetts Ave. NW
Washington, DC 20005
(202) 862-0700

The center analyzes military spending, policies, and weapons systems. CDI opposes the deployment of space weapons, including those of the Strategic Defense Initiative (SDI) and antisatellite weapons. The center publishes *The Defense Monitor* monthly newsletter.

Citizens for Peace in Space
PO Box 915
Colorado Springs, CO 80901
(719) 389-0644

This group, which is dedicated to the peaceful use of outer space, opposes all space weapons. It organizes protests at U.S. military bases that conduct space weapons research. The organization publishes the *Citizens for Peace in Space Newsletter* periodically.

Earth Island Institute
300 Broadway, Suite 28
San Francisco, CA 94133-3312
(415) 788-3666

The institute works to prevent the destruction of the environment. It opposes space weapons and nuclear-powered satellites as threats to the environments of earth and space. Earth Island also opposes rocket launches that deplete earth's ozone layer and any activity that creates debris in outer space. The institute publishes the quarterly *Earth Island Journal*.

Federation of American Scientists (FAS)
307 Massachusetts Ave. NE
Washington, DC 20002
(202) 546-3300

The federation is comprised of scientists, engineers, and individuals concerned with the effects of science on society. FAS opposes the Strategic Defense Initiative and space-based weapons. It publishes the *FAS Public Interest Report* ten times a year.

The Heritage Foundation
214 Massachusetts Ave. NE
Washington, DC 20002
(202) 546-4400

The Heritage Foundation is a public policy research institute that studies many national issues. It supports SDI and a more substantial role for the private sector in space exploration. It publishes research on SDI and NASA in many books and publications, including *The SDI Report* periodic newsletter, *The Heritage Foundation Backgrounder*, *Lectures*, and the quarterly *Policy Review*.

High Frontier
2800 Shirlington Rd., Suite 405A
Arlington, VA 22206
(703) 671-4111

High Frontier pioneered the concept of SDI and is one of the initiative's most ardent supporters. It seeks to use outer space for both military and economic purposes by the United States and its allies. Publications include the quarterly *Journal of Practical Applications in Space* and the monthly newsletter *High Frontier Newswatch*.

Hudson Institute
PO Box 26-919
Indianapolis, IN 46226
(317) 545-1000

The institute studies many national public policy issues, including health, economics, and defense. It advocates a manned mission to Mars, the SDI program, and the development of an economical national aerospace plane (NASP) as an alternative to the space shuttle. The institute opposes construction of space station *Freedom*. Publications include the quarterly *Hudson Institute Report*, the bimonthly journal *Hudson Institute Opinion*, and research papers and books.

Institute for Security and Cooperation in Outer Space (ISCOS)
1336A Corcoran St. NW
Washington, DC 20009
(202) 462-8886

The institute promotes international cooperation and commercial space development, and opposes space-based weapons. ISCOS recommends that SDI be limited to surveillance from space to ensure nations' adherence to peace treaties. Publications include the monthly newsletter *Spaceline*.

National Aeronautics and Space Administration (NASA)
600 Independence Ave. SW
Washington, DC 20546
(202) 453-1000

NASA is the federal agency in charge of the U.S. civil space program. It supports continued use of the space shuttle, construction of space station *Freedom*, and manned exploration of the Moon and Mars. Publications include many books, reports, and the monthly magazine *NASA Activities*.

National Space Council
1600 Pennsylvania Ave. NW
Washington, DC 20500
(202) 456-1414

The council is part of the executive branch of government and is comprised of several cabinet officials, including the vice president. It formulates policy for the U.S. space program and is one of the main advocates of manned missions to the Moon and Mars. The council publishes an annual *Report to the President*.

National Space Society (NSS)
922 Pennsylvania Ave. SE
Washington, DC 20003
(202) 543-1900

The NSS is comprised of individuals who support a strong space program. The organization advocates manned exploration of space, commercial space activities, and space colonization. To achieve such goals, it supports the space station and aerospace plane projects, and missions to the Moon and Mars. The NSS publishes the monthly magazine *Ad Astra*.

The Planetary Society
65 N. Catalina Ave.
Pasadena, CA 91106
(818) 793-5100

The Planetary Society is one of the largest organizations promoting planetary exploration. It advocates both human and robotic space exploration and the search for extraterrestrial intelligence. The society also advocates building a space station that is more practical and less expensive than space station *Freedom*. Publications include *The Planetary Report*, a bimonthly magazine.

SANE/FREEZE: Campaign for Global Security
1819 H St. NW, Suite 1000
Washington, DC 20006
(202) 862-9740

SANE/FREEZE is an advocacy group that seeks an end to nuclear weapons testing and the nuclear arms race. It opposes both space-based and ground-based missile defense programs such as SDI. The organization publishes the quarterly *SANE/FREEZE News*.

The SETI Institute
2035 Landings Dr.
Mountain View, CA 94043
(415) 961-6633

The institute, a research organization partly funded by NASA, is involved in the scientific search for extraterrestrial intelligence. It publishes *The SETI News* quarterly newsletter.

Space Studies Institute (SSI)
PO Box 82
Princeton, NJ 08542
(609) 921-0377

SSI is one of the leading research organizations for outer space colonization. Its research includes studies on lunar materials processing and alternative energy sources in space. Publications include *SSI Update*, a bimonthly newsletter.

Union of Concerned Scientists (UCS)
26 Church St.
Cambridge, MA 02238
(617) 547-5552

UCS is concerned about the impact of advanced technology on society. It opposes the use of weapons in space because it believes such action threatens global peace by extending the arms race into space. UCS publishes a quarterly newsletter, *Nucleus*, as well as books, reports, and briefing papers.

United Nations' Committee on the Peaceful Uses of Outer Space (UN-COPUOS)
United Nations
New York, NY 10017
(212) 963-8302

The committee is composed of UN member nations that cooperate in peaceful space exploration. It meets annually to formulate guidelines for the peaceful use of outer space and reports to the UN general assembly. The committee publishes annual reports, including the *Report of the Committee on the Peaceful Uses of Outer Space*, and other studies.

U.S. General Accounting Office (GAO)
PO Box 6015
Gaithersburg, MD 20877
(202) 275-6241

The GAO, the investigative branch of Congress, is responsible for the examination of all receipts and disbursements of public funds. It testifies before Congress concerning the cost estimates of NASA projects. GAO reports on space-related issues include *Military Space Programs: An Unclassified Overview of Defense Satellite Programs and Launch Activities* and *Questions Remain on the Costs, Uses, and Risks of the Redesigned Space Station*.

Bibliography of Books

Advisory Committee on the Future of the U.S. Space Program
Report of the Advisory Committee on the Future of the U.S. Space Program. Washington, DC: U.S. Government Printing Office, 1990.

Buzz Aldrin and Malcolm McConnell
Men from Earth: An Apollo Astronaut's Exciting Account of America's Space Program. New York: Bantam Books, 1989.

Kent Alexander
Countdown to Glory: NASA's Trials and Triumphs in Space. Los Angeles: Price Stern Sloan, 1989.

Edward Ashpole
The Search for Extraterrestrial Intelligence. New York: Sterling, 1990.

Isaac Asimov
The World's Space Program. Milwaukee: Gareth Stevens, 1990.

William Sims Bainbridge
Goals in Space: American Values and the Future of Technology. Albany, NY: State University of New York Press, 1991.

Mrinal Bali
Space Exploration: A Reference Handbook. Santa Barbara, CA: ABC-CLIO, 1990.

Howard Blum
Out There: The Government's Secret Quest for Extraterrestrials. New York: Simon & Schuster, 1990.

Ben Bova and Byron Preiss, eds.
First Contact: The Search for Extraterrestrial Intelligence. New York: Penguin Books USA, 1990.

Eric Burgess
Return to the Red Planet. New York: Columbia University Press, 1990.

Radford Byerly Jr., ed.
Space Policy Reconsidered. Boulder, CO: Westview Press, 1989.

Philip C. Clarke
The Missile Race: Why We Need Space Defense. Milford, PA: America's Future, Inc., 1990.

Michael Collins
Carrying Fire: An Astronaut's Journey. New York: Bantam Books, 1989.

Michael Collins
Liftoff: The Story of America's Adventure in Space. New York: Grove Press, 1988.

Michael Collins
Mission to Mars. New York: Grove Weidenfeld, 1990.

Congressional Research Service
Outer Space: What Should Be the United States Government Policy Toward the Region Beyond Earth's Atmosphere? Washington, DC: Library of Congress, 1990.

Leonard David
Space Station Freedom: A Foothold on the Future. Washington, DC: NASA, 1988.

George Henry Elias
Breakout into Space: Mission for a Generation. New York: William Morrow, 1990.

Ben R. Finney and Eric M. Jones
Interstellar Migration and the Human Experience. Berkeley: University of California Press, 1985.

Ann Florini
Developing the Final Frontier: International Cooperation in the Peaceful Uses of Outer Space. New York: United Nations Association of the United States of America, 1985.

James J. Frelk and   *Defending Against Ballistic Missile Attacks.* Washing-
Glen E. Tait, eds.   ton, DC: George C. Marshall Institute, 1990.

Louis Friedman   *Starsailing: Solar Sails and Interstellar Travel.* New
York: John Wiley & Sons, 1988.

Jonathan N. Goodrich   *The Commercialization of Space.* New York: Quorum
Books, 1989.

Crockett L. Grabbe   *Space Weapons and the Strategic Defense Initiative.*
Ames: Iowa State University Press, 1991.

David P. Gump   *Space Enterprise: Beyond NASA.* New York: Praeger,
1990.

William K. Hartmann   *Beyond Spaceship Earth.* San Francisco: Sierra Club
Books, 1986.

Kim R. Holmes and   *SDI at the Turning Point: Readying Strategic Defenses for
Baker Spring, eds.   the 1990s and Beyond.* Washington, DC: The Heritage
Foundation, 1990.

Henry Hurt III   *For All Mankind.* New York: The Atlantic Monthly
Press, 1988.

Robert Jastrow   *Journey to the Stars: Space Exploration—Tomorrow and
Beyond.* New York: Bantam Books, 1989.

Linda T. Krug   *Presidential Perspectives on Space Exploration: Guiding
Metaphors from Eisenhower to Bush.* New York:
Praeger, 1991.

Kenneth R. Lang and   *Wanderers in Space: Exploration and Discovery in the
Charles A. Whitney   Solar System.* New York: Cambridge University Press,
1991.

Richard S. Lewis   *Space in the Twenty-First Century.* New York: Columbia
University Press, 1990.

Kenneth N. Luongo and   *The Search for Security in Space.* Ithaca, NY: Cornell
W. Thomas Wander   University Press, 1989.

Howard E. McCurdy   *The Space Station Decision: Incremental Politics and
Technological Choice.* Baltimore, MD: The Johns Hop-
kins University Press, 1990.

Thomas R. McDonough   *The Search for Extraterrestrial Intelligence: Listening for
Life in the Cosmos.* New York: John Wiley & Sons,
1987.

Walter A. McDougall   *The Heavens and the Earth: A Political History of the
Space Age.* New York: Basic Books, 1985.

John L. McLucas   *Space Commerce.* Cambridge, MA: Harvard University Press, 1991.

Jill Steele Mayer, ed.   *Space: The Next Renaissance.* San Diego, CA: Univelt,
1991.

Frank Miles and   *Race to Mars: The Mars Flight Atlas.* New York:
Nicholas Booth   Harper & Row, 1988.

James R. Morrison   *A More Effective Civil Space Program.* Washington, DC:
Center for Strategic and International Studies, 1989.

Bruce Murray   *Journey into Space: The First Thirty Years of Space Exploration.* New York: W.W. Norton and Co., 1989.

National Aeronautics   *Mission to Planet Earth.* Washington, DC: NASA, 1989.
and Space Administration

National Research Council	*Mission to Planet Earth*. Washington, DC: National Academy Press, 1988.
National Space Council	*1991 Report to the President*. Washington, DC: U.S. Government Printing Office, 1991.
Bill Nelson with Jamie Buckingham	*Mission: An American Congressman's Voyage to Space*. San Diego, CA: Harcourt Brace Jovanovich, 1988.
Alcestis R. Oberg	*Spacefarers of the Eighties and Nineties: The Next Thousand People in Space*. New York: Columbia University Press, 1985.
James E. and Alcestis R. Oberg	*Pioneering Space: Living on the Next Frontier*. New York: McGraw-Hill, 1986.
Gerard K. O'Neill	*The High Frontier: Human Colonies in Space*. Garden City, NY: Anchor Books, 1982.
Thomas O. Paine, ed.	*Leaving the Cradle: Human Exploration of Space in the Twenty-First Century*. San Diego, CA: Univelt, 1991.
Robert Powers	*Mars: Our Future on the Red Planet*. Boston: Houghton Mifflin, 1986.
James T. Quinlivan, George L. Donohue, and Edward R. Harshberger	*Strategic Defense Issues for the 1990s*. Santa Monica, CA: Rand Corporation, 1990.
Sally K. Ride	*Leadership and America's Future in Space*. Washington, DC: U.S. Government Printing Office, 1987.
Harry L. Shipman	*Humans in Space: Twenty-First Century Frontiers*. New York: Plenum Press, 1989.
Jeffrey Simon, ed.	*Security Implications of SDI: Will We Be More Secure in 2010?* Washington, DC: National Defense University Press, 1990.
Robert W. Smith	*The Space Telescope: A Study of NASA, Science, Technology, and Politics*. New York: Cambridge University Press, 1989.
Paul B. Stares	*The Militarization of Space: U.S. Policy, 1945-1984*. Ithaca, NY: Cornell University Press, 1985.
Strategic Defense Initiative Organization	*A Guide to International Participation in the Strategic Defense Initiative*. Washington, DC: Strategic Defense Initiative Organization, 1991.
David W. Swift	*SETI Pioneers: Scientists Talk About Their Search for Extraterrestrial Life*. Tucson: The University of Arizona Press, 1990.
Joseph J. Trento	*Prescription for Disaster: From the Glory of Apollo to the Betrayal of the Shuttle*. New York: Crown, 1987.
U.S. Office of Technology Assessment	*Exploring the Moon and Mars: Choices for the Nation*. Washington, DC: U.S. Government Printing Office, 1991.
Wernher von Braun and Frederick I. Ordway III	*Space Travel: A History*. 2nd ed. New York: Harper & Row, 1985.
Frank White	*The Overview Effect*. Boston: Houghton Mifflin, 1987.
Frank White	*The SETI Factor*. New York: Walker and Co., 1990.
John Noble Wilford	*Mars Beckons*. New York: Alfred A. Knopf, 1990.

187

Index

189